Mathematics for level-2 technicians

C. W. Schofield

BSc, FIMA, FCSI, FRSH
Highlands College

and

D. Smethurst

BSc, MSc
Paddington College

Edward Arnold

© C. W. Schofield, D. Smethurst 1979

First published 1979
by Edward Arnold (Publishers) Ltd
41 Bedford Square, London WC1B 3DQ

British Library Cataloguing in Publication Data

Schofield, Clarence Ward
    Mathematics for level-2 technicians.
    1. Shop mathematics
    I. Title II. Smethurst, D
    510'.2'46   TJ1165

    ISBN 0-7131-3385-6

Text set in 10/11 pt IBM Press Roman, printed by photolithography, and
bound in Great Britain, at The Pitman Press, Bath

Contents

Preface

In writing this book, we have aimed to provide a comprehensive student text which extends the fundamentals of mathematics covered in *Basic mathematics for technicians* by C. W. Schofield and relates mathematical theory to a wide range of practical work appropriate to engineering, construction, and science technicians.

We have based the content on the current level-2 standard units in mathematics issued by the Technician Education Council, and have also taken note of the new bank of objectives in mathematics (U78/911) developed by the joint TEC/BEC Committee for Mathematics and Statistics.

Our approach has been greatly influenced by the general availability of electronic calculators which, besides making much detail of arithmetical 'working out' unnecessary, has made an introductory section on calculators desirable, with references to their use in evaluations throughout.

Both of us wish to record our grateful thanks to Bob Davenport of Edward Arnold (Publishers) Ltd, whose careful editing and helpful suggestions have been responsible for many improvements. We are also indebted to many of our colleagues, whose willing assistance we have greatly appreciated.

C. W. Schofield
D. Smethurst

Key to TEC objectives

TEC unit and sections of this book

General objective	U75/012	U75/038	U75/039	U76/031	U76/032	U76/033	U76/060	U76/340
1	D5, D7, D8	D6	D6, D7, D8	B5, B6	C1	B3, B7	A1	*
2	D6	D7	D3	B6, B8	C2	B3	B1, B3	*
3	D1	D8	B3, B7	B3	B8	D2	B5, B7, D2	B4
4	D2	B7	C1	B7	E1, E2	B6, B7	B9	B4
5	E1, E2	B2, B5, B6	C2	E1		B8	D3	D1, D2
6	B5	D3	E1, E2	D5, D7		C1	D7	A1
7	B5, B6	D1, D4				B8, A1		B1, B3, B7
8		E1, E2				A1, A2		B2
9		A1				E1		B1
10						E3		B1
11						D5, D7, D8		B5, B6
12						D8		D5, D8
13						D6		D7
14						B10		D6
15								D5
16								D7, D9

*Revision of level-1 work covered in *Basic mathematics for technicians* by C. W. Schofield.

A   Number

A1 Using a calculator

A1.1 Basic operations

With modern electronic calculators, all arithmetical calculations become simple press-button operations. Numbers are entered by pressing the appropriate button for each digit in the number. By international agreement, the digits from 1 to 9 are arranged in a 3 x 3 block with $\boxed{7}\boxed{8}\boxed{9}$ on the top row, $\boxed{4}\boxed{5}\boxed{6}$ on the centre row, and $\boxed{1}\boxed{2}\boxed{3}$ on the lowest row, with the zero $\boxed{0}$ either to the left or below. Simple calculators may have little more on the remaining buttons than the four basic operations of $\boxed{\times}\boxed{\div}\boxed{+}\boxed{-}$ together with an equals $\boxed{=}$, a cancel or clear $\boxed{C}$ and either a decimal point $\boxed{.}$ or an exponent $\boxed{E}$.

The operating procedure varies according to the type of calculator, and you should refer to the instruction book to find the sequence appropriate to your particular machine.

You should practise using your calculator until you can operate it quickly and accurately. Try using it to check results obtained by using four-figure tables or a slide rule.

Warnings
1. Before starting any calculation, clear the machine (including any memory).
2. Limit your answers to a reasonable number of significant figures consistent with the accuracy of the input data.
3. If using a battery model, switch off your machine between calculations (to conserve battery life).
4. If the display begins to fade, recharge or replace the batteries or the calculations may be affected.

A1.2 Special function keys

Scientific calculators provide a range of special function keys, and in selecting a calculator it is necessary to choose one which has those functions most likely to be needed. Calculators vary in the way these functions are identified and used. Some of the more common key symbols are given below.

$\boxed{\%}$	A percentage key is useful for financial calculations but has limited application in scientific work.
$\boxed{\pi}$	Gives the value of the constant 'pi' to whatever accuracy the register will allow (e.g. 3.141 592 7).

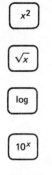

x^2 Pressing this key automatically squares whatever number has been entered.

$\sqrt{x}$ This square-root key is essential for many mathematical calculations.

log The common-logarithm key gives logarithms to base 10.

10^x This key raises 10 to the power of the number which has been entered. It can be regarded as a key for antilogarithms, i.e. the reverse of 'log'.

ln Pressing this key causes the natural logarithm of the entered number to be computed and displayed.

e^x This key raises the exponential e (approximately 2.718 281 8) to the power of the number which has been entered. It thus provides natural antilogarithms, but has other uses since e occurs in many formulae

sin cos tan These keys produce the appropriate trigonometric function for the angle entered. Some calculators are limited to angles from $0°$ to $90°$ but most have a greater range.

arc With calculators with this key, inverse trigonometrical functions are obtained by following this key with the appropriate sine, cosine, or tangent key.

sin^{-1} cos^{-1} tan^{-1} These keys provide an alternative way of obtaining inverse trigonometrical functions. Thus either arcsin x or $sin^{-1} x$ can be used to obtain the angle of which the sine is x.

INV This more general 'INVERSE' key can be used to replace the 'arc' key when inverse trigonometrical functions are required, but it can also be used to give antilogarithms from the logarithm keys and may have other uses.

2

$1/x$	This reciprocal key gives the value of $1 \div x$. It can also be used in combination with the keys for the three basic trigonometrical ratios to give cosecant, secant, and cotangent.
x^y or y^x	This power key raises the first number entered to the power of a second number.
$\sqrt[x]{y}$ or $y^{1/x}$	This root key enables the xth root of any positive value of y to be found.
EXP or EE	This exponent entry key enables numbers to be entered in scientific notation, i.e. in the standard form $x \times 10^n$. The number 0.000123 is displayed as $1.23 - 04$.
$+/-$	This key is used for changing the sign of the number entered.
$x!$	This factorial key gives the value of $x(x-1)(x-2) \ldots (1)$ for integral values of x.
STO STO 1	The number in the display is stored in a cleared memory when this key is used. Some calculators have more than one memory location, and the user can specify which memory register a particular number is to be stored in.
M+	Pressing this key adds the displayed number to the memory.
M-	Subtracts the display from the memory.
RCL or RM	The recall-memory key — content of the memory store to be displayed.
CM	This key clears the memory, and it is advisable to develop a habit of using it every time a fresh calculation is begun.
$x \leftrightarrow M$	Exchanges the displayed number for the number stored in the memory.

3

[(and)]

Keys for brackets or parentheses are very useful in the evaluation of complicated formulae. They permit storage of intermediate results while working out subsections of the calculation.

A1.3 Multiple-function keys

A pocket calculator will rarely have room for more than about forty keys if they are to be reasonably spaced and easy to use. To keep down the number of keys, many manufacturers of scientific calculators allocate second or even third functions to several of the keys. A special 'function' key (often marked 'F' or '2nd') is provided to select which of the uses of a multifunction key is intended. For example, the functions e^x and 10^x may share a single key. Pressing this key will give e^x for the number in the display directly. To find 10^x the 'function' key is pressed first. For example,

1.25 $\boxed{e^x}$ displays 3.4903   i.e. $e^{1.25} = 3.4903$

1.25 $\boxed{F}\boxed{e^x}$ displays 17.7828 i.e. $10^{1.25} = 17.7828$

The second-function symbols are usually marked on the body of the calculator next to the appropriate key, while the first-function symbol is marked on the key itself. Many mistakes are made by inadvertently obtaining the wrong function with these keys.

A1.4 Algebraic entry

Most calculators today operate in a way which allows algebraic functions to be entered in a logical sequence similar to the order in which we would write them down on paper. For example, the problem of adding 8 to 12 and subtracting 6 is normally written as

$$8 + 12 - 6 = 14$$

and we enter it as

8 $\boxed{+}$ 12 $\boxed{-}$ 6 $\boxed{=}$

The answer 14 will appear in the display as soon as we press the 'equals' key.

A1.5 Sequence of operations

Certain calculator functions have precedence over others and will be completed as soon as the function key is pressed. For example, if we clear the calculator and press the sequence of keys

2 $\boxed{x}$ 3 $\boxed{\ln}$

the display will show 1.0986. This is the value of ln 3 and *not* the value of ln (2 x 3). Pressing the 'equals' key will now give 2.1972, which is 2 ln 3. Other functions on the calculator such as 'sin', 'cos', 'tan', 'log', 'e^x', '$1/x$', and 'x^2' also operate immediately on the number in the display regardless of other operations previously entered and not yet carried out.

The system of precedence of functions is called a *hierarchy*, and you should check the calculator handbook to see how your machine operates in this respect.

A1.6 Parentheses and memories

Calculators vary in the way they evaluate a chain of operations involving +, −, x, and ÷ functions. To find (3 x 2) + (4 x 5) = 26, for example, may be a straightforward task on a calculator using the *sums-of-products* precedence. On such a machine we simply use the key sequence

$$3 \boxed{\times} 2 \boxed{+} 4 \boxed{\times} 5 \boxed{=} 26$$

Many other machines will give an answer of 50 with this sequence, i.e. [(3 x 2) + 4] x 5. To obtain the correct value using a machine of this type we use parentheses or the memory as follows:

$$3 \boxed{\times} 2 \boxed{+} \boxed{(} 4 \boxed{\times} 5 \boxed{)} \boxed{=} 26$$

or $\quad 3 \boxed{\times} 2 \boxed{=} \boxed{\text{STO 1}} 4 \boxed{\times} 5 \boxed{=} \boxed{+} \boxed{\text{RCL 1}} \boxed{=} 26$

As a general rule, to avoid mistakes it is advisable to complete separately the evaluation of any expressions within parentheses, just as we would do to solve the problem manually.

A1.7 Checking

Some sort of check is always advisable and, depending on the type of problem, one of the following three ways of checking should be possible.

i) Use amended figures to simplify the calculation and provide a rough guide to the magnitude of the answer to be expected. Thus, in dividing 20.78 by 4.09 we note that 20 divided by 4 would give us 5, so we expect an answer around this figure. Using the calculator gives us 5.08.

ii) Use an alternative mechanism for the calculation, such as four-figure tables or a slide rule. This is good practice at this stage as well as a means of gaining a reasonably accurate answer.

iii) Use an alternative sequence for the calculation on the machine, or reverse it and work back from the answer to the original input. Thus if we have found $(1.36)^4 = 3.421$ we should be able to show that $\sqrt[4]{3.421} = 1.36$.

A1.8 Significant figures

When using a calculator, there is a temptation to read answers from the display and to copy the figures down in full on the assumption that what is shown on the display is accurate to the number of figures produced by the calculator. To illustrate the danger, let us consider a practical example in which we wish to determine the area of a rectangle drawn on a map and measured as 20 mm x 25 mm, given that the accuracy of measurement was to the nearest millimetre. This implies that the dimensions actually lie between

19.5 mm x 24.5 mm and 20.5 mm x 25.5 mm, giving an area somewhere between 477.75 mm^2 and 522.75 mm^2. This shows that the actual area is better expressed as 500 ± 23 mm^2 or as 500 mm^2 ± 4$\frac{1}{2}$%. Now this example is rather an extreme case, but it does illustrate that the accuracy of any final answer is dependent upon the accuracy of the input data. If the reading on an ammeter or the volume of liquid delivered by a burette can be determined with an accuracy of the order of 1%, then any subsequent calculation based upon such readings cannot be taken to have a greater accuracy.

It is apparent from such instances that it would be absurd to perform a calculation based on practical data and then give an answer to as many as eight significant figures (which is the display capacity of many pocket calculators). In most cases it would be realistic to limit such an answer to three significant figures, bearing in mind that the third cannot be guaranteed. Where a more accurate figure is essential, the accuracy of the input data must be increased. This may mean, for example, that instead of measuring a tube diameter by caliper methods we may have to use a travelling microscope to obtain the required accuracy. It is generally desirable to limit any final answer to the same number of significant figures as the input data, but as many figures as possible should be retained throughout the intermediate stages of a calculation.

A1.9 Evaluations
Your calculator will help you in the arithmetical calculations arising from your work in other technical subjects and also in evaluation from formulae.

Example   Find the value of $4\pi r^2$ when $r = 0.39$ m.

Most calculators will allow the evaluation of this expression using a logical sequence of operations from left to right:

		Display
i)	enter 4	4
ii)	press the multiplication key	4
iii)	press the π key	3.141 592 7
iv)	press the multiplication key	12.566 370 6
v)	enter the value of r	0.39
vi)	square using the 'x^2' key	0.1521
vii)	press the '=' key to obtain the result	1.911 345 0

The answer 1.9 m^2 is written down, corrected to the number of significant figures given for the initial value of r (two significant figures). Note that it is necessary to add the appropriate units to the numerical answer.

A1.10 Tabulating values
When it is necessary to make repeated evaluations from the same formula, it is usually helpful to draw up a table showing how the values obtained from the formula relate to the values selected for the variable.

Example 1 Draw up a table of values of y corresponding to values of x from 0 to +3 at intervals of 0.05 for the equation $y = 0.78\sqrt{x} + 0.42$.

The sequence of operations on the calculator is

$$x \boxed{\sqrt{x}} \boxed{\times} .78 \boxed{+} .42 \boxed{=}$$

i.e. i) enter the next value of x
ii) press the square-root key
iii) press the multiplication key
iv) press the decimal key followed by numbers 7 and 8
v) press the addition key
vi) press the decimal key followed by numbers 4 and 2
vii) press the 'equals' key to obtain the result.

The following table of results is obtained:

x	0	0.50	1.00	1.50	2.00	2.50	3.00
y	0.42	0.97	1.20	1.38	1.52	1.65	1.77

Some calculators are *programmable*, which means that a sequence of operations like that above can be stored in the calculator and the program followed through automatically for each value of x entered in turn.

Example 2 Evaluate Ae^{bx} when $x = 1, 4, 10$, given that $A = 1.6$ and $b = \frac{1}{5}$.

For $1.6e^{x/5}$ the calculator sequence is

$$x \boxed{\div} 5 \boxed{=} e^x \boxed{\times} 1.6 \boxed{=}$$

Substituting $x = 1$ yields 1.95

$x = 4$ yields 3.65

$x = 10$ yields 11.82

A1.11 Flow charts
Flow charts are used extensively by computer programmers, who need to establish the best sequence of operations to complete a task. Before writing a program for the computer, they break down the calculation into small steps and arrange them into a suitable order on a diagram, using the following symbols:

         Used to show a terminal point, i.e. start and stop.

          Used when arithmetical operations are to be carried out.

7

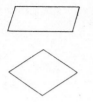 This symbol denotes input or output of numbers.

A decision symbol, determining which alternative path is to be followed.

Symbols are connected by flow lines to indicate the sequence of operations An arrow on the flow line shows the direction.

Flow charts can be used to prepare a problem for calculator solution, particularly when repeated calculations are necessary and a programmable calculator is being used.

Example A graph of the curve $y = 2x^3 - x^2 + x$ is to be drawn for values of x between $x = 0$ and $x = 3$ in steps of 0.2. Draw a flow chart to show how a table of suitable y values could be obtained from a calculator.

The flow chart is shown in fig. A1.1.

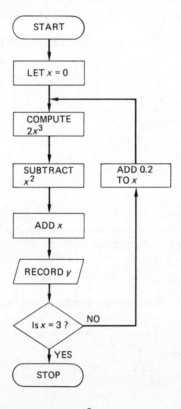

Fig. A1.1

8

Exercise A1

Use a calculator to evaluate each of the following expressions:

1 $3972 + 4165 + 2971 + 385$
2 $2.763 + 0.512 + 11.37 + 7.961$
3 $4734 + 5296 - 3488$
4 $10.356 - 1.983 - 2.497$
5 283×471
6 $2.735 \times 0.625 \times 1.024$
7 $74.35 \div 11.59$
8 $(6.594)^2$
9 $1/2.151$
10 $47.1 \times (0.763)^2$
11 5% of £7360
12 2% of $375
13 $\dfrac{21.73 \times 4.812}{37.16}$
14 $(2.594)^2 - (1.762)^2$
15 $(0.968)^3$
16 $(1.6754)^4$
17 $\dfrac{41.78}{2.421} - \dfrac{0.736}{0.566}$
18 $1/(0.877)^2$
19 $e^{0.2}$
20 $e^{-0.16}$
21 Find the value of πr^2 when $r = 0.518$.
22 Evaluate $2\pi\sqrt{(l/g)}$ when $l = 3.46$ and $g = 9.81$.
23 Given that $1/f = 1/u + 1/v$, find the value of f if $u = 40$ and $v = 160$.
24 If $V = \frac{4}{3}\pi r^3$, find V when $r = 2.4$.
25 Evaluate $\sqrt{s(s-a)(s-b)(s-c)}$ given that $s = \frac{1}{2}(a + b + c)$ when $a = 4.2$, $b = 3.4, c = 2.4$.
26 Using the formula $\dfrac{1}{R} = \dfrac{1}{r_1} + \dfrac{1}{r_2} + \dfrac{1}{r_3}$, find R when $r_1 = 0.820, r_2 = 1.05$, and $r_3 = 1.76$.
27 Use the formula $N = \sqrt{\dfrac{(15D)^5 H}{L}}$ to find the value of N when $D = 1.58$, $H = 51.2, L = 3.4$.
28 Use a calculator to evaluate

  a) $(1.61 \times 2.45) + (3.57 \times 4.11)$
  b) $(11.33 + 2.66) \times (2.18 + 9.67)$
  c) $\dfrac{3.22}{(6.1 \times 1.33)} - \dfrac{2.7}{(8.55 \times 0.32)}$
  d) $\dfrac{(14.62 - 10.26)}{(9.03 + 1.17)}$

29 Draw a flow chart to show how the voltage v across a capacitor increases as t varies between 0 s and 100 s, finding values of v every 5 seconds from the formula $v = E(1 - e^{-t/RC})$, where E, R, and C are constants.

A2 Naperian logarithms

These are logarithms with base e (where e = 2.7183). They are called Naperian or natural logarithms to distinguish them from common logarithms with base 10.

A2.1 Defining logarithms

We define the logarithm of a to the base b as follows:

if $\quad a = b^x \quad$ then $\quad x = \log_b a$

Thus

if $\quad 25 = e^x \quad$ then $\quad x = \log_e 25$

Conversely, if we are given a logarithmic relationship we can use the same definition to obtain an indicial equation. For example

$$11 = \log_x 8$$
$$x^{11} = 8$$

Note that, since $b^0 = 1, \log_b 1 = 0$; and, since $b^1 = b, \log_b b = 1$.

A2.2 Laws of logarithms

i) $\quad \log_b MN = \log_b M + \log_b N$

We can show this is true by using our definition of a logarithm.

Let $\quad \log_b M = x \quad$ and $\quad \log_b N = y$

then $\qquad b^x = M \quad$ and $\qquad b^y = N$

$\therefore \qquad MN = b^x \times b^y = b^{x+y}$

$\therefore \quad \log_b MN = x + y = \log_b M + \log_b N$

ii) $\quad \log_b \dfrac{M}{N} = \log_b M - \log_b N$

Again we let $\quad \log_b M = x \quad$ and $\quad \log_b N = y$

then $\quad \dfrac{M}{N} = \dfrac{b^x}{b^y} = b^{x-y}$

From the definition of a logarithm,

$$\log_b \frac{M}{N} = x - y = \log_b M - \log_b N$$

iii) $\quad \log_b N^a = a \log_b N$

Let $\quad \log_b N = x \quad$ and so $\quad N = b^x$

Raising both sides to the power a,

$$N^a = (b^x)^a = b^{ax}$$

Again using the definition of a logarithm,

$$\log_b N^a = ax = a \log_b N$$

A2.3 Evaluating Naperian logarithms

Although logarithms can be evaluated with any base, the bases 10 and e are most widely used and logarithms with these bases are available in tables and from calculators.

For the equation $y = 10^x$, taking logarithms gives $\log_{10} y = x$; from $y = e^x$ we get $\log_e y = x$.

We shall use $\ln y$ for $\log_e y$ and $\log y$ for $\log_{10} y$ from now on.

Example 1 Find the value of $\ln 60$ (a) from tables, (b) by calculator.

a) Tables usually list the values of $\ln x$ for values of x from 1 to 10. To find the value of $\ln 60$ we write

$$\ln 60 = \ln(6 \times 10)$$
$$= \ln 6 + \ln 10$$
$$= 1.7918 + 2.3026$$
$$= 4.0944$$

b) By calculator, we enter 10 and press 'ln' to get 4.0943.

Example 2 Find the value of $\ln 0.000\,376$ (a) from tables, (b) by calculator.

a) $\ln 0.000\,376 = \ln(3.76 \times 10^{-4})$
$$= \ln 3.76 - 4 \ln 10$$
$$= 1.3244 - 4(2.3026)$$
$$= 1.3244 - 9.2104$$
$$= -7.886 \text{ (or } \overline{8}.114)$$

b) By calculator,

$$\ln 0.000\,376 = -7.8858$$

Since e^x is always positive, it follows that for any number y which is real and positive there exists a number x such that $y = e^x$, i.e. $x = \ln y$.

Example 3 Find x if $e^x = 8.4511$.

$$e^x = 8.4511$$
$$\therefore \quad x = \ln 8.4511$$

11

Using tables or the 'ln' key on a calculator,

$$x = 2.1343$$

Example 4   Find y if $\ln y = 1.8320$.

$$\ln y = 1.8320$$

$$\therefore \quad y = e^{1.8320}$$

From tables or using the 'e^x' key on a calculator,

$$y = 6.2464$$

A2.4   To convert from common to Naperian logarithms
If we know $\log N$ and we wish to find $\ln N$, let $\ln N = x$ so that $N = e^x$.
Now, taking common logarithms of both sides gives

$$\log N = \log e^x = x \log e = \ln N \times \log e$$

$$\therefore \quad \ln N = \frac{\log N}{\log e} = \frac{\log N}{\log 2.7183}$$

$$= \frac{\log N}{0.4343} = 2.3026 \log N$$

i.e. to find the Naperian logarithm of a number we simply multiply the common logarithm of the number by 2.3026.

A2.5 To convert from Naperian to common logarithms
If we know $\ln N$ and we wish to find $\log N$, let $\log N = x$ so that $N = 10^x$.
Taking Naperian logarithms of both sides,

$$\ln N = \ln 10^x = x \ln 10 = \log N \times \ln 10$$

$$\therefore \quad \log N = \frac{\ln N}{\ln 10} = \frac{\ln N}{2.3026} = 0.4343 \ln N$$

i.e. to find the common logarithm of a number we multiply the Naperian logarithm of the number by 0.4343.

Example   If $\log x = 1.8734$ and $\ln y = 2.2241$, find $\ln x$ and $\log y$.

$$\ln x = 2.3026 \log x = 2.3026 \times 1.8734 = 4.3137$$

$$\log y = 0.4343 \ln y = 0.4343 \times 2.2241 = 0.9659$$

A2.6 Equations involving Naperian logarithms
Because laws of natural growth and decay involve exponential functions, it is sometimes necessary to solve equations which involve such functions or their corresponding logarithms.

Example The mass m of a radioactive substance varies with time t according to the equation $m = Ae^{bt}$. Given exactly 5 g of a radioactive substance with a half-life of 63 days, how much of it will be left after 2 weeks?

When $t = 0, m = 5$ g.

We know the substance has a half-life of 63 days, i.e. only half of the original substance will be left after 63 days, so $t = 63$ days when $m = 2.5$ g.

Substituting the two sets of values in $m = Ae^{bt}$,

$$5 = Ae^0$$

but $\quad e^0 = 1$

$\therefore \qquad A = 5$

$$2.5 = 5e^{63b}$$

$\therefore \qquad e^{63b} = 0.5000$

$$63b = \ln 0.5000 = -0.693\,14$$

$\therefore \qquad b = -0.0110$

Hence the equation is $m = 5e^{-0.011t}$.

After 14 days, i.e. $t = 14$,

$$m = 5e^{-0.011(14)} = 5e^{-0.154}$$

$$= 5 \times 0.8572 = 4.286$$

i.e. the mass remaining after 2 weeks will be 4.286 g.

Exercise A2

1 Use a calculator or tables to evaluate each of the following: (a) $\ln 8$, (b) $5 \ln 2$, (c) $1.4 \ln 2.15$, (d) $\ln 0.0078$.
2 Solve the equation $e^{3x} = 7.28$.
3 (a) Find the value of $A(1 - e^{-\omega t})$ when $A = 50$, $\omega = 0.012$, and $t = 4.2$.
  b) Find the value of t if $50(1 - e^{-0.012t}) = 16$.
4 Solve the equation $6.3 \ln x = 1.26$.
5 The voltage v across a capacitor C as it discharges through a resistor R is given by $v = 5e^{-t/RC}$. Find the value of C in microfarads if $v = 1.51$ V at $t = 0.6$ s, given that $R = 10$ kΩ.
6 The mass m of a radioactive substance remaining after t hours is given by the equation $m = Ae^{bt}$, where A is the original mass and b is a constant. If the mass is reduced by half in 184 hours, find the time taken for 10 g of the substance to be reduced to 8 g.

13

B Algebra and graphs

B1 Transposition of formulae

B1.1 Component parts of an equation

An equation represents a balance between two expressions on either side of an 'equals' sign. Mathematicians, scientists, and engineers spend much of their time forming and solving equations. Often there is a single symbol or *variable* on the left-hand side (l.h.s.) and we call this the *subject* of the equation.

On the right-hand side (r.h.s.) may be an expression involving several *terms*. For example, the equation $x = 2y + 3z$ has a subject x on the l.h.s. and two terms on the r.h.s. In cases where a symbol is common to several terms, we say it is a *factor*: as in the equation

$$P = 2QR - QS$$

where Q is a factor of both terms on the r.h.s. We can rewrite this equation as

$$P = Q(2R - S)$$

by taking the factor Q outside a bracket. To regain the original form we simply multiply each term in the bracket by Q.

B1.2 Operations connecting parts of an expression

There are several ways in which the component parts of an expression can be related. The mathematical relationships include addition, subtraction, multiplication, and division. For example, the equations $x = y + z$, $x = y - z$, $x = yz$, and $x = y/z$ all relate the variable x to an expression involving y and z. Other relationships involving powers and roots also occur, as shown in the following examples: $x = y^2 + z^2$, $x = \sqrt{(y + z)}$ $x = y^z$, and $x = \sqrt[z]{y}$.

When performing mathematical operations on equations, we must take care that we maintain the balance implied by the 'equals' sign. If we add a quantity to one side of an equation we must add the same quantity to the other; and this rule applies to all the other operations listed above.

It is often necessary to change the subject of an equation by altering the position of terms, a process we call *transposition*. The rules for the transposition of formulae are basically the same as the rules for solving simple equations, and we shall now consider some of them.

B1.3 Terms connected by a plus or minus sign

It is quite obvious that if $x + 3 = 5$ then $x = 2$, but how do we arrive at this answer? Logically we should proceed as follows.

Write down the original equation: $x + 3 = 5$

Subtract three from each side: $x + 3 - 3 = 5 - 3$

Simplify to $x = 2$

Now the same result could have been obtained simply by transferring the +3 to the other side of the equation and changing the sign to −3, but the justification for this lies in the process outlined above.

Let us consider the equation $y - 1 = 7$ and compare the two methods.

$y - 1 = 7$ or $y - 1 = 7$

Add 1 to each side: Take the −1 across and
                                                   change its sign:
$y - 1 + 1 = 7 + 1$

$\therefore \qquad y = 8$ $y = 7 + 1$

                                          $\therefore \qquad y = 8$

The working for the second method is shorter, and for this reason it is the method usually adopted.

Example   Make T the subject of the equation $N = G - 3T + M$.

First we isolate the term containing T:

$3T = G + M - N$

Now divide both sides by 3:

$$T = \frac{G + M - N}{3}$$

B1.4 A single term involving multiplication or division
The formula relating the area A of a triangle to the base length b and the perpendicular height h is

$$A = \frac{bh}{2}$$

We can transpose this to make b the subject by multiplying both sides by $2/h$, giving

$$A \times \frac{2}{h} = \frac{b\cancel{h}}{\cancel{2}} \times \frac{\cancel{2}}{\cancel{h}}$$

$$\therefore \qquad b = \frac{2A}{h}$$

In this way, any variable which is involved in only one term as a product or fraction (quotient) can be isolated by multiplying or dividing both sides of the the equation by the same amount.

Example Express the radius r of a circle in terms of the circumference c, if $c = 2\pi r$.

$$c = 2\pi r$$

Dividing both sides by 2π gives

$$\frac{c}{2\pi} = r \quad \text{or} \quad r = \frac{c}{2\pi}$$

B1.5 More than one term
When the new subject of a formula occurs in an expression involving more than one term, we must take additional care to treat both sides of the equation in the same way. For example, to make t the subject of the equation

$$v = u + \frac{ft}{m}$$

we first isolate the term in which t occurs:

$$\frac{ft}{m} = v - u$$

Multiplying both sides by m/f gives

$$t = \frac{m}{f}(v - u)$$

Note the use of brackets to express the r.h.s. as a single term instead of $mv/f - mu/f$. This *factorised* form is usually preferred.

Example 1 The force P which is required to pull a weight W up an inclined plane is given by $P = W \sin \alpha + R$. Make W the subject.

$$P = W \sin \alpha + R$$

Isolating the term which includes W,

$$W \sin \alpha = P - R$$

$$\therefore \qquad W = \frac{P - R}{\sin \alpha}$$

Sometimes it takes several operations to isolate a variable and make it the subject of an equation.

Example 2 If $u = v + \frac{W}{5}\left(\frac{xy}{2} - 3z\right)$, make x the subject.

Isolating the term containing x,

$$\frac{W}{5}\left(\frac{xy}{2} - 3z\right) = u - v$$

16

Multiplying both sides by $5/W$,

$$\frac{xy}{2} - 3z = \frac{5}{W}(u - v)$$

or
$$\frac{xy}{2} = \frac{5}{W}(u - v) + 3z$$

Multiplying both sides by $2/y$ gives

$$x = \frac{2}{y}\left[\frac{5}{W}(u - v) + 3z\right]$$

B1.6 Equations involving roots or powers

The presence of a root or power in an equation we wish to transpose should not lead to any additional difficulty provided we always treat both sides of the equation in the same way. To make y the subject of the equation

$$x^2 = 4y^2 - z$$

we isolate the term containing y as before:

$$4y^2 = x^2 + z$$

or
$$y^2 = \frac{x^2 + 2}{4}$$

Taking the square root of both sides,

$$y = \sqrt{\frac{x^2 + z}{4}} = \frac{\sqrt{(x^2 + z)}}{2}$$

Note that $\sqrt{(x^2 + z)}$ does *not* equal $x + \sqrt{z}$.

Example 1 The impedance of an electrical circuit is given by the formula $Z = \sqrt{(R^2 + X^2)}$. Transpose to make X the subject.

$$Z = \sqrt{(R^2 + X^2)}$$

Squaring both sides gives

$$Z^2 = R^2 + X^2$$

$$\therefore \quad X^2 = Z^2 - R^2$$

Taking the square root of both sides gives

$$X = \sqrt{(Z^2 - R^2)}$$

Example 2 Make L the subject of the formula

$$f = \frac{2Md}{(d^2 - L^2)^2}$$

17

Multiplying both sides by $(d^2 - L^2)^2/f$ gives

$$(d^2 - L^2)^2 = \frac{2Md}{f}$$

Taking the square root of both sides,

$$d^2 - L^2 = \sqrt{\frac{2Md}{f}}$$

$$\therefore \qquad L^2 = d^2 - \sqrt{\frac{2Md}{f}}$$

Taking the square root of both sides again,

$$L = \sqrt{\left[d^2 - \sqrt{\left(\frac{2Md}{f} \right)} \right]}$$

B1.7 The new subject is contained in more than one term
When the same variable appears more than once in an equation and we wish to make that variable the subject, we first collect together all terms containing the variable.

To make z the subject of the equation

$$x = \frac{z}{1 + z}$$

multiply both sides by $1 + z$:

$$x(1 + z) = z$$

multiply out the l.h.s.:

$$x + xz = z$$

collect terms involving z:

$$z(1 - x) = x$$

$$\therefore \qquad z = \frac{x}{1 - x}$$

Example Make t the subject of the formula

$$\cos \theta = \frac{1 - t^2}{1 + t^2}$$

Multiplying both sides by $1 + t^2$ gives

$$(1 + t^2) \cos \theta = 1 - t^2$$

$$\therefore \quad \cos \theta + t^2 \cos \theta = 1 - t^2$$

Collecting terms involving t,

$$t^2 (\cos \theta + 1) = 1 - \cos \theta$$

$$\therefore \qquad t^2 = \frac{1 - \cos \theta}{1 + \cos \theta}$$

Taking the square root of both sides,

$$t = \sqrt{\frac{(1 - \cos \theta)}{(1 + \cos \theta)}}$$

B1.8 Substituting given values into formulae

When we are given values of all the variables in a formula except the subject, we can substitute and calculate the value of the subject. Using a calculator can save much time in such calculations.

Example 1   If $w = x \left(1 + \dfrac{yz}{100}\right)$, find w when $x = 550$, $y = 4$ and $z = 30$.

Substituting given values,

$$w = 550 \left(1 + \frac{4 \times 30}{100}\right)$$

$$= 550 (1 + 1.2)$$

$$\therefore \quad w = 1210$$

Some calculators have an operating system which will allow direct entry of such a calculation, including brackets, while others require careful organisation of the order in which operations are carried out. You should ensure that you are familiar with the capability of any calculator you use in this respect.

When the only unknown variable in a formula is not the subject it is usually better to transpose the formula, using methods described in previous sections of this chapter, before substituting numerical values.

Example 2   Calculate M in the formula

$$x = \frac{Ma + mb}{M + m}$$

given that $x = 6.14$, $m = 31.2$, $a = 8.93$, and $b = 4.88$.

$$x = \frac{Ma + mb}{M + m}$$

Transposing to make M the subject,

$$Mx + mx = Ma + mb$$

19

$$\therefore \quad M(x-a) = m(b-x)$$

$$M = \frac{m(b-x)}{(x-a)}$$

Substituting the given values in the r.h.s.,

$$M = \frac{31.2\,(4.88-6.14)}{(6.14-8.93)}$$

$$\therefore \quad M = 14.1$$

When solving practical problems of this type the units used must be consistent, and particular attention should be paid when multiples or submultiples of the SI base units are involved. Quantities measured in kilowatts and millimetres, for example, must first be converted to watts and metres before calculations are carried out.

Example 3 Calculate the heat generated (H) when a current of 250 mA is maintained through a 1.5 kΩ resistor for 30 minutes, if

$$H = I^2Rt \text{ (joules)}$$

For the value of H to be in joules, it is essential to use consistent units for I, R, and t in the formula. These are amperes, ohms, and seconds respectively; therefore

$$I = 250 \times 10^{-3}A \qquad R = 1.5 \times 10^3\,\Omega \qquad t = 30 \times 60\,s$$

Substituting these values gives

$$H = 0.250 \times 1500 \times 1800\,J$$

$$= 675\,000\,J$$

$$= 675\,kJ$$

Exercise B1
In questions 1–5, transpose the given equation to make the variable in brackets to the right of the equation the new subject.

1   a) $z = x + 4y$ [x]

    b) $Q = R - 3S + 4T$ [S]

    c) $b = a - \dfrac{c}{2} - 5d$ [c]

2   a) $V = IR$ [R]

    b) $p_1V_1 = p_2V_2$ [p_2]

    c) $M = \dfrac{fI}{y}$ [y]

d) $I = \dfrac{Prn}{100}$ $\quad\quad$ [r]

e) $E = \dfrac{WL}{Ax}$ $\quad\quad$ [x]

f) $v = u + at$ $\quad\quad$ [t]

3 a) $x = \dfrac{3y + 2z}{5} + w$ $\quad\quad$ [z]

b) $s = ut + \dfrac{Ft^2}{2m}$ $\quad\quad$ [F]

c) $\dfrac{1}{R} = \dfrac{1}{R_1} + \dfrac{1}{R_2}$ $\quad\quad$ [R_1]

d) $\dfrac{x + y}{z} = a - b$ $\quad\quad$ [y]

e) $C = \dfrac{5}{9}(F - 32)$ $\quad\quad$ [F]

f) $a = b + c\left(\dfrac{d}{e} - 2fg\right)$ $\quad\quad$ [d]

g) $p = \dfrac{q}{r + s}\left(t - \dfrac{u + v}{4}\right)$ $\quad\quad$ [u]

4 a) $A = \pi r^2$ $\quad\quad$ [r]

b) $T = 2\pi\sqrt{\dfrac{l}{g}}$ $\quad\quad$ [L]

c) $M = \dfrac{bd^3}{12}$ $\quad\quad$ [d]

d) $d = \sqrt{(b^2 - 4ac)}$ $\quad\quad$ [c]

e) $x = \sqrt{\{y + (k + z)^2\}}$ $\quad\quad$ [z]

f) $J = \left(\dfrac{K + L}{M}\right)^{\frac{3}{2}} - 1$ $\quad\quad$ [L]

5 a) $e = \dfrac{b - a}{b}$ $\quad\quad$ [b]

b) $z = xy + w(1 - 2x)$ $\quad\quad$ [x]

c) $K_1(S_1 - S_2) = K_2S_2$ $\quad\quad$ [S_2]

d) $ax = bx - c(x + d)$ $\quad\quad$ [x]

e) $M = \dfrac{bh^3}{3} - \dfrac{bh^3}{4}$ $\quad\quad$ [h]

f) $P = \left\{ \left(1 - \dfrac{C}{n}\right)^2 + \dfrac{2C}{n} \right\}^2 \quad [n]$

6   The equivalent resistance R of two resistors R_1 and R_2 connected in parallel can be found by using

$$\frac{1}{R} = \frac{1}{R_1} + \frac{1}{R_2}$$

Find R_2 if $R_1 = 2.2\,\mathrm{k\Omega}$ and $R = 387\,\Omega$.

7   The kinetic energy in joules of a moving body is given by $\frac{1}{2}mv^2$, where m is the mass of the body in kg and v is its velocity in m/s. Find the velocity of a body if $m = 800$ g and its kinetic energy is 23.5 kJ.

8   Find the radius of a cone with height $h = 12.5$ mm and volume

$v = 520\,\mathrm{mm}^3$ if $v = \frac{1}{3}\pi r^2 h$.

9   Determine R in ohms in the electrical impedance formula

$$Z = \sqrt{\left[R^2 + \left(\omega L - \frac{1}{\omega C}\right)^2 \right]}$$

when $\omega = 350$ rad/s, $L = 0.4$ H, $C = 6000\,\mu\mathrm{F}$, and $Z = 140\,\Omega$.

B2   Direct and inverse proportionality

B2.1   Dependent and independent variables

In mathematics, when a variable (dependent) can be expressed entirely in terms of another (independent) variable and one or more constants, we say that one variable is a *function* of the other. The circumference c of a circle is a function of the diameter d: we know that $c = \pi d$. We can also write $c = \mathrm{f}(d)$, which means that c is a function of d. Changes in the circle diameter result in corresponding changes in the circumference, and in this case we refer to c as a *dependent* variable while d is the *independent* variable. In the same way, the time period T of a pendulum is a dependent variable while the length of the pendulum l is the independent variable. Again we say T is a function of l or $T = \mathrm{f}(l)$. The relationship in this case is given by

$$T = 2\pi \sqrt{\frac{l}{g}}$$

where g is the acceleration due to gravity and is constant at the place where the pendulum is swinging.

B2.2 Direct proportion

The relationship we looked at in section B2.1 between the diameter d and circumference c of a circle is an example of *direct* proportion. This means that an increase or decrease in d leads to an increase or decrease in c of the same proportion: double the diameter of a circle and the circumference is doubled; halve d and the value of c will be halved. We say that c is directly proportional to d (or that c *varies* directly as d) and write $c \propto d$.

Similarly, in the pendulum example from section B2.1 we observe that the period T is directly proportional to the square root of length l, or $T \propto \sqrt{l}$. This is another example of direct proportion, since T increases as l increases, but we would have to increase the length of l four times to double the period. For example, the lengths of pendulums with periods of one second and two seconds are 248.5 mm and 994.1 mm respectively.

Relationships involving direct proportionality between two or more variables form the basis for many laws in physical science. Here are some examples:

(a) the extension of an elastic body is directly proportional to the applied load (Hooke's law);
(b) the current flowing through a fixed resistor is directly proportional to the applied voltage (Ohm's law);
(c) the volume of a fixed mass of gas at constant pressure is directly proportional to the absolute temperature (Charles's law).

B2.3 Coefficient of proportionality

We saw in section B2.2 that, if y is directly proportion to x, we write $y \propto x$. Consider an example in which $y = 12$ when $x = 4$. Because of the proportionality, the value of y doubles from 12 to 24 when x doubles from 4 to 8. After this change, y is still three times as large as x as it was before. We may replace $y \propto x$ with the equation $y = kx$, where k is a constant usually known as the *coefficient of proportionality*.

In this case, $k = y/x = 12/4 = 3$. Once we have found the coefficient in this way we can calculate y for any new value of x.

Example 1   If $0.2\,\text{m}^3$ of cast iron have a mass of 1380 kg, find the mass of $0.5\,\text{m}^3$ of the same material.

We know that mass m is directly proportional to volume V,

i.e. $m \propto V$ or $m = kV$

If $m = 1380\,\text{kg}$ when $V = 0.2\,\text{m}^3$,

$$k = \frac{m}{V} = \frac{1380}{0.2} = 6900\,\text{kg/m}^3$$

Now, substituting $V = 0.5\,\text{m}^3$ gives

$$m = 0.5 \times 6900 = 3450\,\text{kg}$$

In this example, the coefficient of proportionality is the density of the cast iron.

Example 2 The power P watts in an electrical circuit is directly proportional to the square of the current I flowing in the circuit. Find the change in power from a value of 125 W at $I = 1.8$ A if the current is increased by 0.35 A.

We are told that $P \propto I^2$

$\therefore \quad P = kI^2$

When $I = 1.8$ A, $P = 125$ W

$\therefore \quad 125 = k \times 1.8^2$

$\qquad k = 38.58$

Following the increase in current, $I = 1.8 + 0.35 = 2.15$ A

$\therefore \quad P = kI^2 = 38.58 \times 2.15^2$

$\qquad = 178.34$ W

i.e. there is an increase in power of 53.34 W.

B2.4 Inverse proportion
The electrical resistance R ohms of a length of wire is dependent upon the cross-sectional area A. Resistance and area are not directly proportional, however, since R increases as A *decreases*. In this case, resistance is *inversely proportional* to area (or resistance varies inversely as area). We can write $R \propto 1/A$. If A is doubled, R will be halved, and reducing the area by any factor will increase the wire's resistance by the same factor.

We can introduce a coefficient of proportionality k as we did when dealing with direct proportion in section B2.3:

if $\quad y \propto \dfrac{1}{x} \quad$ then $\quad y = \dfrac{k}{x}$

Rearranging this we see that $xy = k$. In words, this implies that the product of variables which are inversely proportional to each other is a constant. The value of k can be determined from known data and then used to calculate new values for the variables.

Example 1 Boyle's law tells us that at constant temperature the volume of a fixed mass of gas is inversely proportional to the pressure upon it. Air in a cylinder occupying a volume of 0.008 m³ at 5 bars pressure is compressed by a piston until the volume is reduced by 0.0015 m³. Determine the final pressure.

From Boyle's law,

$\qquad V \propto 1/p \quad$ or $\quad V = k/p$

Substituting the initial values of V and p,

$$0.008 = \frac{k}{5}$$

$\therefore \qquad k = 0.040$

After compression, the volume of air is $0.0065 \, \text{m}^3$, but

$$p = \frac{k}{V}$$

$$= \frac{0.040}{0.0065} = 6.154 \text{ bars}$$

i.e. the final pressure is 6.15 bars.

Example 2   Figure B2.1 shows a variable pulley system used in a vehicle transmission. Pulley A has a fixed diameter and is driven at a constant speed by the engine. The diameter of pulley B can be changed, and, when it is 150 mm, B rotates at 1000 rev/min. If the speed of rotation of B is inversely proportional to its diameter, find the coefficient of proportionality and the change in diameter necessary to bring the speed of B down to 800 rev/min.

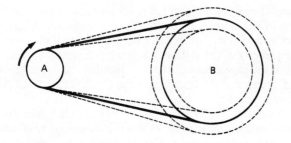

Fig. B2.1 Variable pulley system

Let the speed of B be N and its diameter be d; then

$$N \propto \frac{1}{d} \quad \text{or} \quad N = \frac{k}{d}$$

When $N = 1000 \text{ rev/min}, d = 150 \text{ mm}$

$\therefore \quad k = Nd = 1000 \times 150$

$\qquad = 150\,000$

To reduce the speed of B to $N = 800 \text{ rev/min}$,

$$800 = \frac{150\,000}{d}$$

$$\therefore \qquad d = 187.5$$

i.e. the diameter must be increased by 37.5 mm.

Exercise B2
1   In each of the two cases below, express one variable as a function of the other, stating which is the dependent and which the independent variable:

a) the length L of a steel rod and the temperature θ;
b) the distance D between two towns and the time t to walk between them.

2   State what is meant when we say that one variable is in direct proportion to another, and show how we express this mathematically.
3   Evaluate the constants of proportionality in the following examples:

a) if y is directly proportional to x and $y = 1.75$ when $x = 0.7$;
b) if y varies directly as the square root of x and $y = 81$ when $x = 9$;
c) if y is directly proportional to the cube of x and $y = 15$ when $x = 2.6$.

4   Hooke's law states that for an elastic body the extension produced is directly proportional to the load applied. A spring has a coefficient of proportionality (or stiffness) $k = 200$ N/m. Find the load required to extend it by 30 mm.
5   The gravitational force of attraction F between two bodies varies inversely as the square of the distance D between them. Express this relationship as a proportionality and in the form of an equation involving a constant k.
6   The frequency f of the voltage in a circuit is inversely proportional to the square root of the capacitance C. If $f = 5$ Hz when the capacitance $C = 2000\ \mu$F, determine the frequency when $C = 2500\ \mu$F. What is the coefficient of proportionality?
7   The shutter speed for correct exposure on a camera is inversely proportional to the square of the aperture, which is indicated by an 'f' number. Explain why standard apertures of f2.8, f4, and f5.6 are used.
8   The second moment of area I for a steel bar of constant width is directly proportional to the cube of the depth D of the bar. Find the coefficient of proportionality if $I = 845$ mm^4 for a bar which is 25 mm deep. What would be the second moment of area of a bar with the same width but twice the depth?

B3   Quadratic equations

B3.1   Expressions and equations
A *quadratic expression* in x has the form $ax^2 + bx + c$; for example:

$$2x^2 + 3x + 4 \quad \text{where } a = 2, b = 3, \text{ and } c = 4$$

26

$$3x^2 + 6 \qquad\qquad a = 3, b = 0, \text{and } c = 6$$

$$4x - x^2 \qquad\qquad a = -1, b = 4, \text{and } c = 0$$

Note that all these expressions involve x^2, or 'x in the second degree'. A *quadratic equation* is obtained by making a quadratic expression equal to zero and has the form $ax^2 + bx + c = 0$; for example:

$$8x^2 - 3x + 2 = 0 \quad \text{where } a = 8, b = -3, c = 2$$

$$3x^2 + x = 4 \qquad\qquad a = 3, b = 1, \quad c = -4$$

$$6.5 - 2.3x^2 - 5x = 0 \qquad\qquad a = -2.3, b = -5, c = 6.5$$

B3.2 Roots of a quadratic equation

Quadratic equations have two *solutions*. This means that there are two values of x which when substituted into the equation will make the quadratic expression on one side equal to zero. These solutions are called *roots* of the equation, and finding the roots is called *solving* the equation. If we substitute $x = 1$ or $x = 2$ in the equation $x^2 - 3x + 2 = 0$, the l.h.s. is zero in both cases. The roots of this equation are therefore $x = 1$ and $x = 2$. There are no other values of x which make the l.h.s. equal to zero.

A graph of $y = x^2 - 3x + 2$ would show that this curve crosses the x-axis at the points $x = 1$ and $x = 2$. This is to be expected, since $y = 0$ along the x-axis, and we know that these two values of x satisfy the equation $x^2 - 3x + 2 = 0$. Roots are therefore the values of x at which a curve drawn from the quadratic equation crosses the x-axis. We can in fact determine roots graphically in this way, but such a method will take more time and will be less accurate than other analytical methods we shall consider. When a quadratic curve does not cross the x-axis, we say it has no *real* roots.

B3.3 Solution by factors

To solve a quadratic equation by factors, we make use of the fact that, if the product of two quantities is zero, then one or the other (or both) of the quantities must be zero.

If $m \times n = 0$ then $m = 0$ or $n = 0$. The same reasoning applies if $(m - 2)(n - 3) = 0$: we can say that $m - 2 = 0$ or $n - 3 = 0$, giving $m = 2$ or $n = 3$. Many quadratic equations can be broken into factors, or *factorised*, in this way to give the roots.

If we take as an example the equation $x^2 - 3x + 2 = 0$, which we saw in section B3.2 has roots $x = 1$ and $x = 2$, we can rewrite it in the factorised form $(x - 1)(x - 2) = 0$. Multiplying out these brackets confirms that it is the same equation.

To factorise the quadratic equation $2x^2 + x - 6 = 0$, we first draw a pair of brackets and equate to zero.

$$(\qquad)(\qquad) = 0$$

The term $2x^2$ has factors $2x$ and x, and we can put these in place:

$$(2x \quad)(x \quad) = 0$$

Now consider the constant term; $6 = 6 \times 1$ or 2×3, and there are four ways we could arrange these numbers in the brackets:

$$(2x \pm 6)(x \pm 1)$$
$$(2x \pm 1)(x \pm 6)$$
$$(2x \pm 2)(x \pm 3)$$
$$(2x \pm 3)(x \pm 2)$$

Only one arrangement with the correct signs will give the correct equation when multiplied out. Clearly the factors involving 6 and 1 will not give the x term required in the original equation, and after some thought we see that the factors are

$$(2x - 3)(x + 2) = 0$$

Check by multiplying out: $2x^2 + 4x - 3x - 6 = 0$, or $2x^2 + x - 6 = 0$. This is the original equation and therefore the factors are correct. The roots of the equation can now be deduced.

If $\quad (2x - 3)(x + 2) = 0$

then $\quad 2x - 3 = 0 \quad$ or $\quad x + 2 = 0$

$\therefore \qquad x = 3/2 \quad$ or $\quad x = -2$

The roots are $x = 3/2$ and $x = -2$.

Example Factorise to find the roots of $3x^2 + 11x = 4$.

The quadratic equation is

$$3x^2 + 11x - 4 = 0$$

Factorising gives

$$(3x - 1)(x + 4) = 0$$

Checking by multiplication,

$$3x^2 + 12x - x - 4 = 0 \quad \text{or} \quad 3x^2 + 11x - 4 = 0$$

The roots are $x = \frac{1}{3}$ and $x = -4$.

B3.4 Deducing a quadratic equation from its roots
We have seen in section B3.3 how to factorise a quadratic expression to obtain the roots of an equation. If the factorised form of an equation is $(x - p)(x - q) = 0$, then $x = p$ and $x = q$ are the roots. It follows that, if we are given the roots, it is always possible to write down factors and then

28

multiply to obtain the equation. If, for example, $x = 3$ and $x = 4$ are the roots of a quadratic equation, we first write down factors and equate to zero:

$$(x - 3)(x - 4) = 0$$

(Note the minus signs here; if the root is negative we use a plus sign in the factor.) Multiplying these brackets gives

$$x^2 - 4x - 3x + 12 = 0$$

or $\qquad x^2 - 7x + 12 = 0$

This is the quadratic equation of which the roots are $x = 3$ and $x = 4$.

Example　Find the quadratic equation of which the roots are $x = \frac{1}{2}$ and $x = -\frac{3}{4}$.

From the roots, $\quad x = \frac{1}{2} \quad$ or $\quad 2x = 1$

and $\qquad\qquad\quad x = -\frac{3}{4} \quad$ or $\quad 4x = -3$

The factors are therefore

$$(2x - 1)(4x + 3) = 0$$

Multiplying these gives

$$8x^2 + 6x - 4x - 3 = 0$$

The quadratic equation is therefore

$$8x^2 + 2x - 3 = 0$$

Some quadratic equations have equal roots, which may be positive or negative. Consider an equation with equal roots $x = 5$. The factorised form of this is

$$(x - 5)(x - 5) = 0$$

This is a *perfect square* and can be written as

$$(x - 5)^2 = 0$$

The quadratic equation is found by multiplication of the factors as before:

$$x^2 - 10x + 25 = 0$$

When the roots of a quadratic equation have the same magnitude but different signs, for example $x = 3$ and $x = -3$, we find that the quadratic equation has no term in x; since

$$(x - 3)(x + 3) = 0$$

giving $\quad x^2 + 3x - 3x - 9 = 0$

or $\qquad\qquad\quad x^2 - 9 = 0$

(the *difference of two squares*).

B3.5 Solutions by formula

We can find the roots of any quadratic equation $ax^2 + bx + c = 0$ by substituting values of a, b, and c into the formula

$$x = \frac{-b \pm \sqrt{(b^2 - 4ac)}}{2a}$$

Two roots are obtained by first adding and then subtracting the value of the square root in the formula. For example, to find the roots of the equation

$$2x^2 + x - 6 = 0$$

we substitute $a = 2$, $b = 1$, and $c = -6$, giving

$$x = \frac{-1 \pm \sqrt{[1 - 4 \times 2 \times (-6)]}}{4} = \frac{-1 \pm \sqrt{49}}{4}$$

The roots are therefore

$$x = \frac{-1 + 7}{4} = \frac{3}{2} \quad \text{and} \quad x = \frac{-1 - 7}{4} = -2$$

These are, of course, the same values we found by factorising this equation in section B3.3.

Example 1 Find the roots of the quadratic equation $3.2x^2 + 1.4x - 7.6 = 0$ using the formula

$$x = \frac{-b \pm \sqrt{(b^2 - 4ac)}}{2a}$$

Substituting $a = 3.2$, $b = 1.4$, and $c = -7.6$ gives

$$x = \frac{-1.4 \pm \sqrt{[(1.4)^2 - 4 \times 3.2 \times (-7.6)]}}{2 \times 3.2}$$

$$= \frac{-1.4 \pm \sqrt{(1.96 + 97.28)}}{6.4} = \frac{-1.4 \pm \sqrt{99.24}}{6.4}$$

$$\therefore \quad x = \frac{-1.4 + 9.96}{6.4} \quad \text{or} \quad x = \frac{-1.4 - 9.96}{6.4}$$

The roots are $x = 1.34$ and $x = -1.78$.

The quadratic formula may be derived by making part of the general quadratic equation $ax^2 + bx + c = 0$ into a perfect square. We know that

$$(x + p)^2 = x^2 + p^2 + 2xp$$

(the square of the first term plus the square of the second term together with twice the product). Consider the equation

$$ax^2 + bx + c = 0$$

30

or $\qquad ax^2 + bx = -c$

$$x^2 + \frac{b}{a}x = -\frac{c}{a}$$

We make the l.h.s. of this equation into a perfect square by adding $(b/2a)^2$ to both sides:

$$x^2 + \frac{b}{a}x + \left(\frac{b}{2a}\right)^2 = -\frac{c}{a} + \left(\frac{b}{2a}\right)^2$$

or $\qquad \left(x + \frac{b}{2a}\right)^2 = \frac{b^2 - 4ac}{4a^2}$

Taking the square root of both sides,

$$x + \frac{b}{2a} = \pm\sqrt{\frac{b^2 - 4ac}{4a^2}}$$

giving $\qquad x = \frac{-b \pm \sqrt{(b^2 - 4ac)}}{2a}$

This formula should be remembered, as it provides a way of finding the roots of quadratic equations which we cannot factorise easily.

There are three types of solution possible, depending on the nature of the square root in the formula. If $b^2 - 4ac$ is positive there will be two different roots of the quadratic equation. When $b^2 = 4ac$ the equation has equal roots. If $b^2 - 4ac$ is negative, the evaluation of roots is beyond the scope of this unit and we say the equation has no *real* roots. For a graphical interpretation of these three cases see fig. B3.1. Note that the graph of a quadratic equation is parabolic.

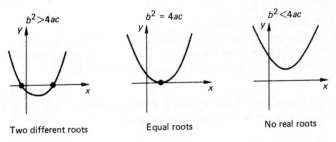

$b^2 > 4ac$

$b^2 = 4ac$

$b^2 < 4ac$

Two different roots

Equal roots

No real roots

Fig. B3.1 Roots of the equation $ax^2 + bx + c = 0$

Example 2   For what value of k will the equation $kx^2 + 6x + 3 = 0$ have equal roots?

The general quadratic equation $ax^2 + bx + c = 0$ has equal roots when $b^2 = 4ac$. Substituting $a = k$, $b = 6$, and $c = 3$, there will be equal roots when

31

$$36 = 4 \times k \times 3$$

or $\quad k = \dfrac{36}{12} = 3$

Check: the equation $3x^2 + 6x + 3 = 0$ will factorise to give $(3x + 3)(x + 1) = 0$ and has equal roots $x = -1$.

B3.6 Practical problems leading to quadratic equations

Example 1 The height s of an object at time t after being thrown vertically up with a velocity u is given by the equation $s = ut - \frac{1}{2}gt^2$. Find t when the object is 6 m high, given that $u = 18$ m/s and $g = 9.81$ m/s^2.

$$s = ut - \tfrac{1}{2}gt^2$$

Substituting the values given,

$$6 = 18t - 4.91t^2$$

or $\quad 4.91t^2 - 18t + 6 = 0$

Using the formula to solve this quadratic equation,

$$t = \frac{-b \pm \sqrt{(b^2 - 4ac)}}{2a}$$

where $\quad a = 4.91 \qquad b = -18 \quad$ and $\quad c = 6$

$$\therefore \quad t = \frac{18 \pm \sqrt{(324 - 117.84)}}{9.81} = \frac{18 \pm 14.36}{9.81}$$

giving $t = 3.30$ s and $t = 0.37$ s.

The two answers in this problem indicate that the object first reaches a height of 6 m after 0.37 s on the way up. Then, after reaching maximum height, it falls back and is 6 m high a second time after 3.30 s.

In some problems, one of the roots of a quadratic equation may be of no use, or even physically impossible, and may be ignored.

Example 2 The area of metal in fig. B3.2, which shows a cross-section of an aluminium extrusion, is 2000 mm^2. Find x, the uniform thickness of the walls.

From fig. B3.2, the outside dimensions of the cross-section are $50 + 2x$ and $30 + 2x$. The area of metal can be expressed as follows:

$$2000 = (50 + 2x)(30 + 2x) - 50 \times 30$$

$$= 1500 + 100x + 60x + 4x^2 - 1500$$

32

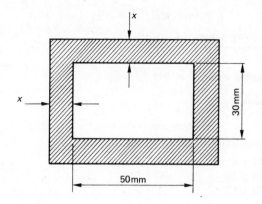

Fig. B3.2

Rewriting this quadratic equation,

$$4x^2 + 160x - 2000 = 0$$

Dividing by 4,

$$x^2 + 40x - 500 = 0$$

Factorising the l.h.s.,

$$(x + 50)(x - 10) = 0$$

The roots are $x = -50$ mm and $x = 10$ mm.

The negative root clearly has no meaning in this case, and the wall thickness of the extrusion is 10 mm.

Example 3 When a current I flows through a resistance R, the power in the circuit is given by I^2R. When I is increased by 3 A with R constant, the power is doubled. Determine the positive value of I.

Let the power before and after the change in current be

$$P_1 = I^2R \quad \text{and} \quad P_2 = (I+3)^2R$$

We know that $P_2 = 2P_1$

$$\therefore \qquad (I+3)^2R = 2I^2R$$

or $(I^2 + 6I + 9)R = 2I^2R$

R will cancel and we have

$$I^2 - 6I - 9 = 0$$

Using the formula,

$$I = \frac{-b \pm \sqrt{(b^2 - 4ac)}}{2a}$$

33

where $\quad a = 1 \qquad b = -6 \quad$ and $\quad c = -9$

$$\therefore \quad I = \frac{6 \pm \sqrt{(36 + 36)}}{2} = \frac{6 \pm 8.49}{2}$$

Taking the positive value,

$$I = 7.24 \text{ A}$$

Exercise B3

1   Factorise the following quadratic expressions:
   a) $x^2 + 5x + 6$
   b) $m^2 + m - 20$
   c) $x^2 - 8x + 12$
   d) $2z^2 + 7z + 3$
   e) $13x + 10 - 3x^2$
   f) $6r^2 - 5r - 4$
   g) $5 - 11x - 12x^2$

2   Use factors to find the roots of the following equations:

   a) $x^2 - 7x + 12 = 0$
   b) $2a^2 - 9a - 18 = 0$
   c) $6 - 7y - 5y^2 = 0$
   d) $3x^2 = 10 - 13x$
   e) $k(5k - 9) = 2$
   f) $x^2 - 6x + 9 = 0$

   g) $t = \dfrac{30}{2t - 7}$

   h) $\dfrac{y + 5}{y - 3} = \dfrac{1}{y}$

3   Use the formula method to solve the following quadratic equations:

   a) $2x^2 + 3x - 4 = 0$
   b) $3.8p^2 + 6.2p - 11.5 = 0$
   c) $y^2 - 17y + 72.25 = 0$

   d) $5 - \dfrac{x}{2} - \dfrac{x^2}{8} = 0$

4   If $t^2 - 4t + a$ is a perfect square, what is a?
5   For what value of k will the equation $kx^2 - 3.5x + 1.9 = 0$ have equal roots?
6   The angle (θ radians) turned through by an accelerating shaft in t seconds is given by

$$\theta = \omega t + \tfrac{1}{2}at^2$$

Find the time taken to complete half a revolution if $\omega = 2.7$ rad/s and $a = 0.8$ rad/s^2.

7   The impedance Z of an electrical circuit is given by $Z = \sqrt{(R^2 + X^2)}$.
Find R if $Z = 27\ \Omega$ when $X = (R + 4)\ \Omega$.

8   If one side of a rectangle is 3 mm longer than the other, and the length of
the diagonals is 10 mm, find the length of the sides.

9   A workshop is 61 m long and 35 m wide. A concrete path surrounds the
workshop. If the area of the path is $400\ \mathrm{m}^2$, find its width.

10   The end of the steel bar in fig. B3.3 is turned on a lathe to a diameter of
44 mm. If the area of the end is now $416\pi\ \mathrm{mm}^2$ less than that of the original
bar, find the diameter of the bar.

Ø 44mm

Fig. B3.3

11   When a cable of length l is stretched between two points a distance x
apart, the sag y in the cable is given by the equation $l = 2.67\ y^2/x + x$. Find
the value of x when $l = 120$ m and $y = 4$ m.

12   For the dissociation of hydrofluoric acid, $k = 6.9 \times 10^{-4}$. Find x when
$k = x^2/(0.100 - x)$.

B4   Indicial equations

B4.1   The laws of indices

Using *index notation* we can write $32 = 2^5$, where 2^5 implies a string of five
twos multiplied together, i.e. $2 \times 2 \times 2 \times 2 \times 2$. In this notation, the lower
number (in large type) is known as the *base* and the upper number (usually in
smaller type) is called the *index* (plural *indices*). 2^5 has base 2 and index 5; it
is usually read as 'two to the power five'. Whole-number values of index are
commonly referred to as *powers* in this way.

Multiplication
Consider the product $2^3 \times 2^2$, which is equal to $(2 \times 2 \times 2) \times (2 \times 2) = 32$.
We can express 32 as 2^5, and so $2^3 \times 2^2 = 2^5$. The index 5 is the sum of the
indices 3 and 2.

In general, when two quantities *with the same base* are multiplied,

$$a^m \times a^n = a^{m+n}$$

Division
We can deduce that

$$\frac{a^m}{a^n} = a^{m-n}$$

Consider the numerical example

$$\frac{2^3}{2^2} = \frac{\cancel{2} \times \cancel{2} \times 2}{\cancel{2} \times \cancel{2}} = 2$$

Now, $2 = 2^1$ and we conclude that

$$\frac{2^3}{2^2} = 2^{3-2} = 2^1$$

Power of a power
The expression $(2^3)^2$ implies that we square the value of 2^3, giving $(2 \times 2 \times 2)$ $\times (2 \times 2 \times 2) = 2^6$. In general,

$$\left(a^m\right)^n = a^{m \times n}$$

i.e. we multiply the indices.

Roots
We define the nth root of a as the number which multiplied by itself n times gives a. We write $\sqrt{a}$ (square root), $\sqrt[3]{a}$ (cube root) and $\sqrt[n]{a}$ (nth root). Note that we omit the 2 in the case of the square root.

Using our definition,

$$\sqrt[3]{a} \times \sqrt[3]{a} \times \sqrt[3]{a} = a$$

We know from the multiplication law of indices that

$$a^{\frac{1}{3}} \times a^{\frac{1}{3}} \times a^{\frac{1}{3}} = a$$

and therefore, in general,

$$\sqrt[n]{a} = a^{1/n}$$

The value of a^0
We have seen that $a^m/a^n = a^{m-n}$ and in cases when $m = n$ we have $a^m/a^m = a^{m-m} = a^0$. It is also clear that, by cancelling, $a^m/a^m = 1$ and so

$$a^0 = 1$$

This is a general rule for any value of a.

Reciprocals in indicial form

Again using the division law for indices, $a^m/a^n = a^{m-n}$, and taking the case when $m = 0$, the numerator becomes $a^0 = 1$ and we have

$$1/a^n = a^{0-n} = a^{-n}$$

This gives the general rule that

$$\frac{1}{a^n} = a^{-n}$$

Example Express $(1/x^3)^2$ in indicial form.

$$\left(\frac{1}{x^3}\right) = x^{-3} \quad \text{and} \quad (x^{-3})^2 = x^{-3\times 2} = x^{-6}$$

B4.2 Change of base

An *indicial* equation is one involving indices, which may be numeric or algebraic, for example $x^3 = 8$ or $3^x = 27$. The six rules of indices listed below can be used to solve simple equations of this type.

Rules of indices

i) $a^m \times a^n = a^{m+n}$

ii) $a^m \div a^n = a^{m-n}$

iii) $(a^m)^n = a^{mn}$

iv) $\sqrt[n]{(a^m)} = a^{m/n}$

v) $\dfrac{1}{a^n} = a^{-n}$

vi) $a^0 = 1$

Rule (iii) is often particularly useful as a first step in the solution.

The expression 8^2 can be written with a new *base*. Since we know that $8 = 2^3$, it follows that

$$8^2 = (2^3)^2$$

The third rule tells us to multiply the indices

$$\therefore \quad 8^2 = 2^6$$

It is sometimes necessary to change the base using a series of steps, and some numbers may be expressed in a variety of ways.

Example Express 729 to bases 9 and 3.

Dividing by 9 gives $729 = 81 \times 9$

and $$81 = 9 \times 9$$

so $$729 = 9^3$$

This is the first answer and we obtain the second by replacing 9 by 3^2, so that

$$9^3 = (3^2)^3$$

The third rule of indices gives

$$(3^2)^3 = 3^6$$

so $$729 = 9^3 \text{ or } 3^6$$

B4.3 Algebraic indices

In section B4.1 we saw how to change the base of a number using the rules of indices. If the index includes unknown variables, the method remains the same. We can express 8^x to the base 2 by writing

$$8^x = (2^3)^x = 2^{3x}$$

Example 1   Express 64^{2x} as 4^{nx}.

Writing 64 as 4^3 we see that

$$64^{2x} = (4^3)^{2x} = 4^{6x}$$

Example 2   Write 27^{2x+1} as an expression with base 3.

We know that $27 = 3^3$

so $$27^{2x+1} = (3^3)^{2x+1} = 3^{3(2x+1)} = 3^{6x+3}$$

(Note that from the first rule in section B4.1 that we could also write this as $3^{6x} \times 3^3$.)

B4.4 Equations with linear indices

We can use the methods in sections B4.1 and B4.2 to solve some types of indicial equations. The method is to find a common base for both sides of the equation and then equate the indices. As a simple example to illustrate the method, suppose we are asked to find x given that $81 = 3^x$. We first write the left-hand side with base 3:

$$3^4 = 3^x$$

$\therefore$ $x = 4$

Example 1   Solve the indicial equation $2^{3x-1} = 32$.

We first express 32 in indicial form with base 2, so that

$$2^{3x-1} = 2^5$$

then, equating indices,

$$3x - 1 = 5$$

$$\therefore \quad 3x = 6$$

$$x = 2$$

Checking this solution by substituting $x = 2$ in the original equation,

$$2^{6-1} = 2^5 = 32$$

Example 2 Solve $9^{x-1} = 27^{2x+1}$.

First change the bases:

$$(3^2)^{x-1} = (3^3)^{2x+1}$$

$$\therefore \quad 3^{2x-2} = 3^{6x+3}$$

Equating indices,

$$2x - 2 = 6x + 3$$

$$\therefore \quad 4x = -5$$

$$x = -\tfrac{5}{4}$$

Check:

$$9^{(-\frac{5}{4})-1} = 0.007 \quad \text{and} \quad 27^{2(-\frac{5}{4})+1} = 0.007$$

(The y^x function on a calculator can be most useful in checking.)

Remember that we can equate indices only when the base is the same on both sides of the equation.

B4.5 Equations with quadratic indices

When the indicial equation involves quadratic indices, the solution will often require us to factorise or use the formula method to obtain two values.

Consider the equation

$$8^{x^2-1} = 64$$

Changing to a common base gives

$$8^{x^2-1} = 8^2$$

Equating indices,

$$x^2 - 1 = 2$$

$$\therefore \quad x^2 = 3$$

and

$$x = \sqrt{3} = \pm 1.732$$

(Note that there are always positive and negative values of a square root.)

Check: $8^{3-1} = 8^2 = 64$ for both values of x.

39

Example Solve the indicial equation $5^{2x^2+1} = 125^x$.

Make the base on both sides the same:

$$5^{2x^2+1} = 5^{3x}$$

Equating indices,

$$2x^2 + 1 = 3x$$

or $\quad 2x^2 - 3x + 1 = 0$

Solving by factorising,

$$(2x - 1)(x - 1) = 0$$

The solutions are

$$2x - 1 = 0$$

$$x = \tfrac{1}{2}$$

and $\quad x - 1 = 0$

$$x = 1$$

Checking both these answers by substituing them in the original equation: if $x = \tfrac{1}{2}$, the l.h.s. becomes $5^{(\frac{2}{4})+1} = 5^{\frac{3}{2}}$ as does the r.h.s., and when $x = 1$ both sides are again the same and equal to 5^3. The solutions are therefore $x = \tfrac{1}{2}$ and $x = 1$.

B4.6 Simultaneous indicial equations

We are already aware that when solving linear simultaneous equations an independent equation is needed for each unknown variable involved. The same rule applies for simultaneous indicial equations with more than one unknown. The technique from sections B4.2 and B4.3 of reducing both sides of an equation to a common base and equating indices can be used to give two or more simultaneous linear equations. These can be solved by the process of elimination and substitution.

Consider the equations

$$2^{x+y} = 4 \quad \text{and} \quad 3^{x-2y} = 27$$

The first step is to change bases on the right of both equations:

$$2^{x+y} = 2^2 \quad \text{and} \quad 3^{x-2y} = 3^3$$

Equating indices in each case gives

$$x + y = 2 \tag{i}$$

$$x - 2y = 3 \tag{ii}$$

Subtracting equation (ii) from equation (i) to eliminate x,

$$3y = -1$$

$$\therefore \quad y = -\tfrac{1}{3}.$$

Next we substitute this value of y in either equation (i) or equation (ii) to find x:

$$x + (-\tfrac{1}{3}) = 2$$
$$x = 2 + \tfrac{1}{3} = \tfrac{7}{3}$$

The solutions $y = -\tfrac{1}{3}$ and $x = \tfrac{7}{3}$ can be checked by substituting into both of the original indicial equations. For example,

$$2^{\frac{7}{3}-\frac{1}{3}} = 2^2 = 4$$

and $\quad 3^{\frac{7}{3}-(-\frac{2}{3})} = 3^3 = 27$

The solutions are thus confirmed.

Example   Solve the equations $4^{2x} = 64^{y+1}$, $5x = 25^{3y-2}$.

Changing bases to make them equal on both sides,

$$4^{2x} = (4^3)^{y+1} = 4^{3y+3}$$
$$5^x = (5^2)^{3y-2} = 5^{6y-4}$$

Equating indices from both equations gives

$$2x = 3y + 3 \tag{i}$$
$$x = 6y - 4 \tag{ii}$$

Multiplying equation (ii) by 2 gives

$$2x = 12y - 8 \tag{iii}$$

Subtracting equation (i) from equation (iii),

$$0 = 9y - 11$$
$$\therefore \quad y = \tfrac{11}{9}$$

Substituting this value of y in equation (i),

$$2x = \tfrac{33}{9} + 3 = \tfrac{60}{9}$$
$$\therefore \quad x = \tfrac{10}{3}$$

Checking these values of x and y by substitution in the original indicial equations using a calculator gives

$$4^{2(\frac{10}{3})} = 10\,321.27 = 64^{\frac{11}{9}+1}$$

and $\quad 5^{\frac{10}{3}} = 213.75 = 25^{\frac{33}{9}-2}$

The solutions are confirmed as $x = \tfrac{10}{3}$ and $y = \tfrac{11}{9}$.

Other methods of solving indicial equations using logarithms are covered in later mathematics units.

Exercise B4

1 Convert each of the following numbers to index notation with the bases required:

    a) 343 base 7
    b) 512 bases 8 and 2
    c) 6561 bases 9 and 3
    d) $5^2 \times 2^4$ base 20

2 Simplify the following expressions:

    (a) $x^3 \times x^2$ (b) $x^4 \times x^{1.5}$ (c) $a^5 \times a^{-3}$

    (d) $x^{(y+1)}x^{(2y-3)}$ (e) b^3/b^2 (f) $y^{1.8}/y^{0.3}$ (g) z^{x-1}/z^{x+2}

3 Express the following in simple indicial form.

    (a) $(x^3)^4$ (b) $(v^{1.2})^{2.3}$ (c) $(a^b)^{-c}$ (d) $\sqrt[3]{x}$

    (e) $\sqrt[x]{y}$ (f) $\sqrt[t]{(s^2)}$ (g) $\sqrt[a]{(b^a)}$

4 Change the base of the following expressions to that required:

    a) 27^x base 3
    b) 64^{2x} bases 4 and 8
    c) $25^{y/2}$ base 5
    d) $\sqrt[x]{81}$ base 3

5 Solve the following indicial equations without using logarithms:

    a) $3^{x+2} = 81$
    b) $25^{2x} = 125$
    c) $36^{z-1} = 6^{2-3z}$
    d) $(\sqrt{3})^y = 9^{y-1}$

6 Determine both solutions to the following equations without using logarithms:

    a) $11^{x^2+1} = 121$
    b) $36^{2-x^2} = 216$
    c) $13^{x^2} = 169^{x+4}$
    d) $7^{z-2} = 343^{1/z}$
    e) $2^y 4^{2y+1} = 16^{y^2}$

7 Solve the simultaneous indicial equations below without using logarithms:

    a) $3^{x-y} = 9^{2x-1}$

        $2^{3y+1} = 8^{x+2y}$

    b) $16^{2x-1} = 64^{x-y}$

        $5^{y-2} = 25^{3x-2}$

B5 Straight-line graphs

B5.1 Plotting a graph

When plotting a graph, it is conventional to use two axes at right angles, inter-secting at the origin O. The horizontal axis OX is usually referred to as the x-axis and the vertical axis OY as the y-axis. The position of any point on the graph is then fixed if its distance from each axis is given, such distances being known as *co-ordinates*.

In fig. B5.1, the point P on the straight-line graph corresponds to the value 4 on the x-axis and to the value 3 on the y-axis, i.e. $y = 3$ when $x = 4$. Writing co-ordinates in the form (x,y), we say that P is the point $(4,3)$. Similarly, point Q has co-ordinates $(10,6)$.

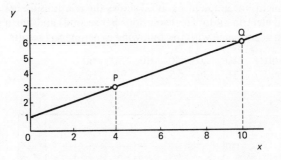

Fig. B5.1

It is advisable to observe the following details:

i) Give the graph a title, to explain what it is all about.

ii) Make full use of the graph paper — the larger the graph the more accurate it is likely to be.

iii) Choose scales which are easy to work from — it is far better to let one unit on the graph paper represent 1, 2, 5, or 10, say, rather than 3, 7, or 9.

iv) It is often possible to start one or both scales from zero, but never do so if it will result in the points being crowded into one small corner of the graph paper.

v) Write enough figures along each axis to help in reading the graph, but avoid overcrowding.

vi) Label each axis to explain what the numbers represent, remembering to put in the units where appropriate.

vii) Mark each point on the graph either by a fine dot on the exact spot (with a circle round it to make it easy to find) or by a cross in the form of a 'plus' sign, with the intersection exactly over the required location.

Either method can be used to indicate the accuracy of an experimental observation. In the first case, the diameter of the circle drawn round á particular point can indicate the possible error limits of measurement. In

43

the second case the extent of possible inaccuracy can be indicated by the lengths of the short horizontal and vertical lines which form the cross and which intersect over the optimum value.

Plotting from experimental data

At this stage we are concerned with straight-line graphs, which are the easiest to draw since we can use a ruler or other straight edge. The line should be drawn in pencil, never with a ball-point or fibre-tip pen. Ideally it should pass through all the points which have been plotted, but we should realise that it does not necessarily pass through the origin as well unless (0,0) is one of the points we have plotted. In experimental graphs in which we have used circles round the points to indicate possible error limits, we should find that our graph line will intersect all such circles if we have deduced the right relationship between the two variables.

The straight-line graph in fig. B5.2 shows the relationship between the speed of an electric motor (N revolutions per second) and the applied voltage (V volts) in accordance with the following experimental results:

V (volts)	10	20	30	40	50	60
N (rev/s)	18	36	55	71	90	108

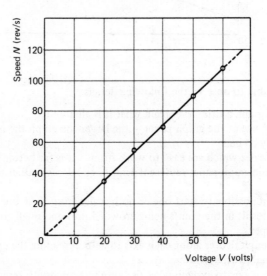

Fig. B5.2 Electric-motor experiment

Having plotted the points and drawn the straight-line graph, we can use it to estimate what speed of rotation will result from a certain voltage, or what voltage will produce a particular speed. If, for example, we want to know what voltage to apply to get a speed of rotation of 45 rev/s, we simply trace across horizontally at the value of 45 on the N-axis until we reach the line of the graph, then drop vertically on to the V-axis to find that the required value is exactly 25 volts.

B5.2 The straight-line graph

Any equation which can be expressed in the form $y = mx + c$ will give a straight-line graph. The quantity in the position of y in the equation is plotted on the vertical axis, and the other variable along the horizontal axis.

For the equation $y = 2x - 3$, we see that when $x = 0, y = -3$; when $x = 2, y = 1$; and when $x = 4, y = 5$. This implies that the three points $(0,-3)$ $(2,1)$ $(4,5)$ should all lie on the same straight line. Figure B5.3 shows that when the three points are plotted they all lie on one straight line, and it follows that this is the graph of the equation $y = 2x - 3$.

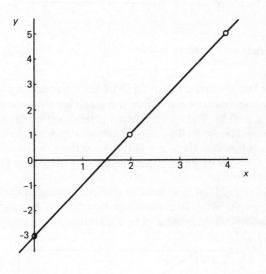

Fig. B5.3

Gradient and intercept

In fig. B5.4, the lower straight-line has the equation $y = 0.5x$ (which could also be written as $y = \frac{1}{2}x$, or $2y = x$, or $x = 2y$). The advantage of writing the equation in the form $y = 0.5x$ is that the coefficient of x (in this case 0.5) gives us the *gradient* or *slope* of the graph.

Consider the section of the graph from O (the origin) to P (the point with co-ordinates $x = 4, y = 2$). By the time it reaches P, the graph has risen 2 units vertically while moving 4 units horizontally. Its gradient or slope is thus $\frac{2}{4} = 0.5$. If we consider the section from O to Q, we find the gradient to be $\frac{3}{6} = 0.5$. The areas between OP and the x-axis and OQ and the x-axis are similar triangles, and the line OPQ makes an angle θ with the x-axis such that $\tan \theta = 0.5$. Using a calculator (or tangent tables), we find that $\theta = 26.57°$ ($26° 34'$).

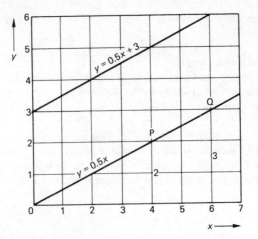

Fig. B5.4 Graphs with positive gradient

The upper straight-line graph in fig. B5.4 has the equation $y = 0.5x + 3$. Here also the coefficient of x is 0.5, so this graph has the same gradient as the lower graph $y = 0.5x$. The two graphs are parallel, having the same gradient.

From the equation $y = 0.5x + 3$, we can see that when $x = 0, y = 3$. The graph therefore cuts the y-axis at $(0,3)$ and the +3 in its equation gives us the distance from the origin at which the line cuts the y-axis. This length is known as the *intercept*.

The graphs in fig. B5.4 have positive gradient, sloping upwards from left to right. The graphs in fig. B5.5 have negative gradient. A straight-line graph with zero gradient would be parallel to the x-axis.

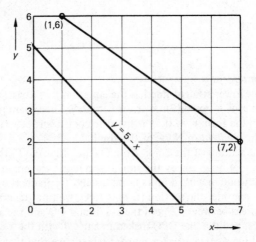

Fig. B5.5 Graphs with negative gradient

46

In fig. B5.5, the lower graph has a gradient of −1 and the intercept on the y-axis is +5, so its equation is $y = 5 - x$.

In general, any straight-line graph will have an equation of the form $y = mx + c$, where m is the gradient and c is the intercept. We can find the equation of the second graph in fig. B5.5 by substituting the known co-ordinates (x and y values) in the equation $y = mx + c$ and solving the resulting pair of simultaneous equations to find the appropriate values of m and c.

Since $y = 6$ when $x = 1$,	$6 = m + c$	
Since $y = 2$ when $x = 7$,	$2 = 7m + c$	
By subtraction,	$4 = -6m$	$m = -\frac{2}{3}$
By substitution,	$6 = -\frac{2}{3} + c$	$c = 6\frac{2}{3}$
So the equation is	$y = -\frac{2}{3}x + 6\frac{2}{3}$	

which may be written as $\qquad 2x + 3y = 20$

B5.3 The determination of laws

We have seen that any equation of the form $y = mx + c$ will give a straight-line graph. The converse of this is also true, i.e., if a given set of values when plotted gives points which (within the limits of experimental accuracy) lie on a straight line, then a relationship of the form $y = mx + c$ must exist.

Example The following results are from an experiment with a set of pulley blocks.

Effort E (newtons)	8	11.9	16	20.2	24	27.9
Load W (newtons)	20	40	60	80	100	120

It is known that these forces should be related by the law $E = aW + b$. If this is so, determine the values of the constants a and b.

Compare $\quad E = aW + b$

with $\qquad y = mx + c$

The variable E is in the position of y and should therefore be plotted along the vertical axis, while W, being in the place of x, will go horizontally. If the resulting graph is a straight line, then its gradient will be a and the intercept on the vertical axis will be b.

The graph is shown in fig. B5.6, and is clearly a straight line. This proves the validity of the relationship $E = aW + b$. The intercept is seen to be 4, therefore $b = 4$. The gradient is $\frac{16}{80} = \frac{1}{5}$, therefore $a = \frac{1}{5}$. Hence the complete relationship is $E = \frac{1}{5}W + 4$.

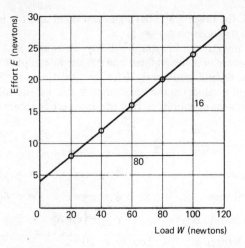

Fig. B5.6 Pulley-block experiment

Once an equation has been determined in this way, the relationship can be used to calculate other values; for example, from the equation $E = \frac{1}{5}W + 4$, we see that when $W = 50$, E will be 14, i.e. an effort of 14 N will be required to lift a load of 50 N. It can be checked from the graph that the point with co-ordinates (50, 14) lies on the straight line.

It should be noted that calculating other values from a law which has been determined in this way is trustworthy only over the range of values for which the relationship is known to hold true. For experimental laws derived from practical experiments, there may come a point at which a physical change occurs that makes the former relationship no longer valid. Examples include stretching of wires, for which extension is proportional to load (Hooke's law) until necking occurs, and the passing of electrical currents through wires until melting occurs. Such limitations do not apply to purely mathematical relationships such as the formulae used in mensuration.

Exercise B5

1 For the equation $y = 3x - 4$, find the values of y corresponding to $x = 1$, $x = 2$, $x = 3$. Plot the points with these co-ordinates and show that they lie on the same straight line.

2 A company has a certain installation initially valued at £12 500. The table below shows the depreciation in the value of this installation. Plot a graph of value against time in years, and from the graph find how many years it will take for the value to drop to one tenth of its initial value.

Time (years)	0	2	6	10	16
Value (£)	12 500	11 250	8750	6250	2500

3 The quantity of a certain chemical which will dissolve in a fixed volume of water at temperature T is given in the table below.

Mass m (kg)	22	23	24	26	28	32
Temperature T (K)	310	315	320	330	340	360

Plot a graph of m against T and verify that they are connected by a law of the form $m = aT + b$. From the graph, find the values of constants a and b.

4 Verify graphically that the relationship $V = E + IR$ is satisfied by the following set of values of V and I.

I (amperes)	0.2	0.5	0.8	1.2	1.5
V (volts)	7.4	11.9	16.4	22.4	26.9

Use the graph to estimate the values of E and R.

5 Without plotting the graphs, find the gradients and intercepts on the y-axis for each of the following:

(a) $2y = 4x + 3$ (b) $y + 5x = 7$ (c) $4y - 6x = 9$

(d) $2x - 8y = 5$ (e) $9x = 5y - 2$ (f) $7 - x - y = 0$

(g) $\dfrac{x + 5}{2} = \dfrac{y - 3}{4}$ (h) $\dfrac{y}{4} - \dfrac{x}{5} = 1$

6 Plot the graphs of the following equations, using only one pair of axes, and thus show that they form a square and find its area.

(a) $y = \dfrac{3(2 - x)}{2}$ (b) $\dfrac{y}{2} = \dfrac{x + 4}{3}$

(c) $3x + 2y + 3 = 0$ (d) $2x = 3y + 1$

7 Draw the straight-line graph through the points $(4, 2)$ and $(1, -3)$. Measure the intercept and gradient, and so find the equation of the straight line. Check by substituting the values in the equation $y = mx + c$ and solving the simultaneous equations for m and c.

8 In an experiment, readings are taken of temperature θ after time t. It is known that θ and t are connected by a relationship $\theta = a + bt$. Given that the temperature is 10°C after 5 seconds and 20°C after 15 seconds, find the constants a and b.

9 A straight-line graph is given by the equation $y = mx + c$. Given that $y = 4$ when $x = 2$, and that $y = 7$ when $x = 8$, find the values of the constants m and c. Hence find the value of x which makes $y = 5$.

10 If stress σ and absolute temperature T are related by the law $\sigma = a + bT$, verify that the following values fit a straight-line graph within error limits of ±2% and hence estimate the values of the constants a and b.

T (K)	280	300	325	350	380	420
σ (Pa)	194	189	185	180	175	166

B6 Non-linear physical laws

The equation $y = ax^2 + c$ would give a curve if we plotted y against x. Although the intercept in this case would enable us to find the value of the constant c, the gradient is not constant and would not give us a directly. We can, however, represent any quantities we choose along the axes, and so the method in this case is to plot y against x^2, taking x^2 as our new variable in place of x. Comparison of

$$y = mx + c$$

and $\quad y = ax^2 + c$

shows that plotting y against x^2 will give a straight-line graph, with gradient a and intercept c.

Many other laws can be reduced to straight-line forms by similar methods, and thus the values of the constants can be determined provided that sets of values of the variables are given. Methods of dealing with the most common types of equations are as follows.

$y = ax^3 + b$, etc.
With equations of the forms $y = ax^2 + b$, $y = ax^3 + b$, $y = a\sqrt{x} + b$, $y = a/x + b$ and $y = a/x^2 + b$, plot y against $x^2, x^3, \sqrt{x}, 1/x$ or $1/x^2$ as appropriate. Comparison with $y = mx + c$ shows that the gradient m will give a, and the intercept c will give the value of b.

$y = ax^n$
Take logs of each side:

$$\log y = \log a + n \log x$$

Compare $\quad y = c + mx$

Plot $\log y$ against $\log x$ to obtain a straight-line graph with gradient n and intercept $\log a$; a can then be found by using a calculator or a table of antilogarithms.

$y = ab^x$
Take logs of each side:

$$\log y = \log a + x \log b$$

Compare $\quad y = c + mx$

Plot $\log y$ against x to obtain a straight-line graph with gradient $\log b$ and intercept $\log a$; use antilogs to find a and b.

$$y = ae^{bx}$$

Take logs of each side:

$$\log y = \log a + bx \log e \qquad (\log_{10}e = 0.4343)$$

Compare $\quad y = c + mx$

Plot $\log y$ against $\log x$ to obtain a straight-line graph with gradient $0.4343b$ and intercept $\log a$.

$$y = ax^2 + bx + c$$

This contains three constants and it is necessary to find the value of one of them before attempting to obtain a straight-line form. The method is to plot a graph of y against x, or at least as much of it as is necessary, to find the intercept on the y axis. This gives the value of c.

Rearrange the equation:

$$y - c = ax^2 + bx$$

Divide by x:

$$\frac{y - c}{x} = ax + b$$

Draw up a set of values for $(y - c)/x$ and plot against x. The gradient will be a and the intercept b.

Other equations containing three constants, e.g. $y = ax^n + b$, can be dealt with similarly.

If the given values are known to be accurate, they can be substituted in the given law and the resulting equations solved simultaneously. When the given values are liable to experimental error, it is better to plot the graph, select two appropriate points on the curve and use their co-ordinates to substitute to give the simultaneous equations. For a law with three constants, three points would be required and three simultaneous equations. An equation such as $y = a\,e^{bx} + c$ could be treated in this way, whereas it could not be put into straight-line form until the value of c had been determined. Note that with this equation when $x = 0$, $y = a + c$, therefore the value of c cannot be found by the intercept on the y-axis.

Example 1 Measurements of electrical power P (in kilowatts) for a range of currents I (amperes) in a heater element are given below. Draw a straight-line graph, plotting P against I^2, to show that $P = I^2R$. Use your graph to find the value of R in ohms.

P (kW)	0	0.10	0.20	0.30	0.40	0.50
I (A)	0	1.30	1.85	2.21	2.61	2.86

Values of I^2 are

	0	1.69	3.42	4.88	6.81	8.18

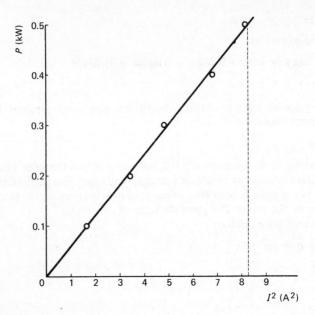

Fig. B6.1

Plotting P against I^2 gives the straight line in fig. B6.1.

The slope of the line is $\dfrac{0.5 \times 1000}{8.3} = 60$

i.e. the equation of the line is $P = 60I^2$ and the resistance of the heater element is 60 Ω.

Example 2   Verify graphically that the table of values below corresponds to a law of the form $y = ax^n$ and, from the graph, determine values of a and n.

x	2.00	4.00	6.00	8.00	10.00
y	7.88	19.40	32.87	47.77	63.85

If $y = ax^n$ then $\log y = n \log x + \log a$ and plotting $\log y$ against $\log x$ will give a straight line. Values of $\log y$ and $\log x$ are as follows:

$\log x$	0.30	0.60	0.78	0.90	1.00
$\log y$	0.90	1.29	1.52	1.68	1.81

The graph drawn using these values is a straight line (see fig. B6.2) and therefore the equation relating y and x is of the form $y = ax^n$. The slope of the line is

$$\frac{1.81 - 0.51}{1.00} = 1.3 \quad \therefore \quad n = 1.3$$

52

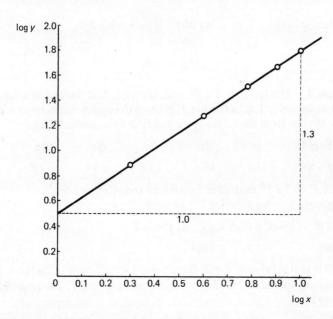

Fig. B6.2 Graph of $\log y = n \log x + \log a$

The intercept on the $\log y$ axis $= 0.51$

$\therefore \quad \log a = 0.51$

and $\quad a = 3.2$

i.e. the equation is $y = 3.2x^{1.3}$.

Example 3 Two variables, x and y, are connected by an equation of the form $y = ax^n$. Given $y = 67.5$ when $x = 3$ and $y = 20$ when $x = 2$, find the value of the constants a and n.

Taking logs, $\qquad \log y = \log a + n \log x$

Substituting values, $\quad \log 67.5 = \log a + n \log 3$ $\qquad$ (i)

$$\log 20 = \log a + n \log 2 \qquad \text{(ii)}$$

The values of the logs can be put in at this stage and the equations worked out with the resulting numbers, or, alternatively,

subtracting, $\qquad \log 67.5 - \log 20 = n(\log 3 - \log 2)$

$$n = \frac{\log 3.375}{\log 1.5} = \frac{0.5283}{0.1761}$$

$$= 3$$

53

substituting in (ii), $\log 20 = \log a + 3 \log 2$

$$20 = 8a$$

$\therefore$ $a = 2.5$

Example 4 The law $F = T\mathrm{e}^{-\mu\theta}$ gives the force F at the end of a rope round a capstan to hold a tension T, θ being the angle of contact in radians. Verify the law from the values given and find the constants T and μ.

θ (rad)	2π	4π	6π	8π	10π
F (N)	170	48.6	13.8	3.94	1.12

The law $F = T\mathrm{e}^{-\mu\theta}$ must first be put into straight-line form.
Taking logs, $\log F = \log T - \mu\theta \log \mathrm{e}$

$$\log F = \log T - 0.4343\,\mu\theta$$

Compare $y = c + mx$

If $\log F$ is plotted against θ, this should give a straight line if the law is true. The gradient should be $-0.4343\,\mu$ and the intercept $\log T$. The required table of values is

$\log F$	2.230	1.687	1.140	0.596	0.049
θ	2π	4π	6π	8π	10π

The graph is shown in fig. B6.3 and, as it is a straight line, the law is verified.

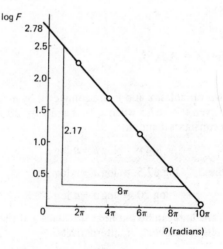

Fig. B6.3 Graph of $\log F = \log T - 0.4343\,\mu\theta$

From the graph, the gradient of which is negative,

$$0.4343\,\mu = 2.17 \div 8\pi$$

$$\therefore \qquad \mu = \frac{2.17}{3.4744\pi} = 0.200\,19$$

and intercept

$$\log T = 2.78$$

$$\therefore \qquad T = 602.56$$

But these values have been worked out to more figures than can be relied upon, and therefore it would be reasonable to state the law as

$$F = 600\,e^{-\theta/5} \text{ approximately}$$

Exercise B6

1 Corresponding values of two quantities are given in the following table:

x	2.0	4.0	6.0	8.0	10.0
y	4.4	6.8	10.8	16.4	23.6

Verify that these quantities follow a law of the form $y = a + bx^2$ by plotting a suitable straight-line graph, and hence find the most probable values of a and b.

2 If the safe loading which can be carried by a girder of a certain design varies as shown in the following table, prove graphically that the load (L) is related to the span (S) by the relationship $L = K/S$, and hence find the value of the constant K.

S(m)	3.0	3.5	4.2	4.5	5.0
L(kN)	315	270	225	210	189

3 The time of oscillation t of a pendulum is believed to be related to the length l by the formula $t = A\sqrt{l}$. The following measurements were taken to an accuracy of ± 2 mm and ± 0.05 seconds.

l (mm)	300	400	500	650	750	850
t (s)	11.0	12.7	14.2	16.2	17.4	18.5

Plot a straight-line graph and use it to find the value of the constant A.

4 Verify graphically that the following table of values corresponds to a law of the form $y = ax^2 + bx$, and hence determine a and b.

x	1	2	3	4	5
y	−6	−7	−3	6	20

5 The results of an experiment on the deflection of a beam are given in the table below. Verify that the deflection (y) and the length of the beam (x) are connected by the relationship $y = ax^n$, and use graphical methods to estimate the values of the constants a and n.

x (m)	1.6	1.8	2.1	2.5	3.2
y (mm)	6.6	10.5	19.5	39.0	104.8

6   Using the following table of values, verify graphically that the head of pressure h and the flow velocity v are connected by the relationship $v = kh^n$, and find the constants k and n.

h	12.2	15.7	19.5	23.4	28.9
v	8.4	9.5	10.5	11.7	12.9

B7 Solving equations graphically

Simultaneous linear equations (i.e. equations giving straight-line graphs) can be easily and speedily solved by algebraic methods, but these methods become more and more involved as the degree of the equations increases, i.e. if there are terms in x^2, x^3, etc. Thus, if one of a set of simultaneous equations is, say, a cubic, then it is usually easier to solve the simultaneous equations by means of a suitable graph or graphs. To illustrate the technique, we will first show how linear equations may be solved graphically.

B7.1 Simultaneous linear equations
If two straight-line graphs are plotted, the co-ordinates of the point of intersection are the required values which satisfy the equations of both graphs.

Example   Solve the simultaneous equations $y = 1 + 0.4x, y = 4 - 0.6x$.

For the first equation, when $x = 0, y = 1.0$

$$\text{when } x = 2, y = 1.8$$

$$\text{when } x = 4, y = 2.6$$

For the second equation,   when $x = 0, y = 4.0$

$$\text{when } x = 2, y = 2.8$$

$$\text{when } x = 4, y = 1.6$$

Plotting these values enables us to draw the two straight-line graphs shown in fig. B7.1.

The graphs intersect where $x = 3$ and $y = 2.2$. Substitution in the original equations shows that these values fit both equations and thus the required solution is $x = 3, y = 2.2$.

Solution of the equations by algebraic methods also confirms these values of x and y, as follows.

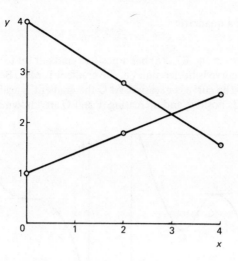

Fig. B7.1

Solution by substitution

Numbering the equations (i) and (ii),

$$y = 1 + 0.4x \qquad \ldots \ldots \ldots \ldots \ldots \ldots \ldots \text{(i)}$$

$$y = 4 - 0.6x \qquad \ldots \ldots \ldots \ldots \ldots \ldots \ldots \text{(ii)}$$

Substituting $y = 1 + 0.4x$ from equation (i) into equation (ii) gives

$$1 + 0.4x = 4 - 0.6x$$

$$\therefore \qquad x = 3$$

Substituting this value for x in equation (i) gives

$$y = 1 + 1.2$$

$$\therefore \qquad y = 2.2$$

Solution by elimination

We can eliminate y from equations (i) and (ii) by subtracting equation (ii) from equation (i) to give

$$0 = -3 + x$$

$$\therefore \qquad x = 3$$

Substituting this value for x in either equation (i) or equation (ii) gives $y = 2.2$.

57

B7.2 Graph of a quadratic

Turning points

The curve shown in fig. B7.2(a) has a positive gradient up to point P. At P the tangent to the curve is horizontal, i.e. the gradient is zero. Between P and Q the gradient of the curve is negative. At Q the gradient is again zero. Beyond Q the gradient is positive and increasing. P and Q are known as *turning points*.

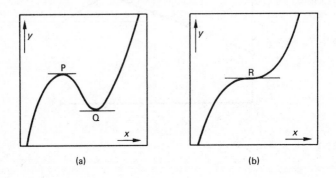

(a)   (b)

Fig. B7.2 Turning points

At P the curve has a *maximum* value. This does not mean that the curve never goes higher than P but rather that P is the highest point on one particular section of the curve. At Q the curve has a *minimum* value.

Now consider the curve shown in fig. B7.2(b). We have seen that with a turning point which is a maximum the gradient changes from positive to negative as x increases up to and past the point. Also, with a turning point which is a minimum the gradient changes from negative to positive as x increases up to and past the point. The curve in fig. B7.2(b) has a tangent to the curve at R which is parallel to the x-axis, i.e. with gradient zero, but on either side of R the gradient is positive. We say that at R we have a *horizontal point of inflexion*.

We can see that, if in fig. B7.2(a) we bring P and Q closer and closer together until the maximum at P coincides with the minimum at Q, the result will be a curve similar to that in fig. B7.2(b). A curve with a horizontal point of inflexion but with the gradient negative on either side will have a similar shape to that in fig. B7.2(b) but the curve will slope in the opposite direction.

The curves shown in figs B7.2(a) and (b) are of cubic equations, having the general form $y = ax^3 + bx^2 + cx + d$ and $y = A(x - B)^3 + C$ respectively, but the graph of a quadratic equation will also have either a maximum or a minimum, as discussed below.

Graph of the general quadratic

The graph of the general quadratic $y = ax^2 + bx + c$ is *parabolic*. Figure B7.3 shows the effect of giving various values to the constants a, b, c.

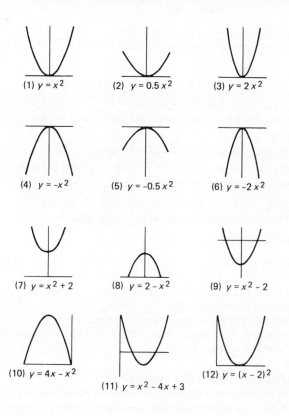

(1) $y = x^2$

(2) $y = 0.5 x^2$

(3) $y = 2 x^2$

(4) $y = -x^2$

(5) $y = -0.5 x^2$

(6) $y = -2 x^2$

(7) $y = x^2 + 2$

(8) $y = 2 - x^2$

(9) $y = x^2 - 2$

(10) $y = 4x - x^2$

(11) $y = x^2 - 4x + 3$

(12) $y = (x - 2)^2$

Fig. B7.3 Graphs of quadratics

1) $a = 1, b = c = 0$; i.e. $y = x^2$
The curve has its minimum at the origin.

2) $a = 0.5, b = c = 0$; i.e. $y = 0.5x^2$
Reducing the value of a flattens the curve.

3) $a = 2, b = c = 0$; i.e. $y = 2x^2$
Increasing the coefficient of x^2 produces a steeper curve.

4) $a = -1, b = c = 0$; i.e. $y = -x^2$
When the coefficient of x^2 is negative, the curve is inverted.

5) $a = -0.5, b = c = 0$; i.e. $y = -0.5x^2$
A flatter curve with its maximum at the origin.

6) $a = -2, b = c = 0$; i.e. $y = -2x^2$
A steeper curve but position of maximum unchanged.

7) $a = 1, b = 0, c = 2$; i.e. $y = x^2 + 2$
The curve is displaced 2 units up the y-axis but is still symmetrical.

8) $a = -1, b = 0, c = 2$; i.e. $y = 2 - x^2$
The curve is inverted and crosses the x-axis at $x = \pm\sqrt{2}$.

9) $a = 1, b = 0, c = -2$; i.e. $y = x^2 - 2$
The intercept on the y-axis is now -2. The roots are $x = \pm\sqrt{2}$.

10) $a = -1, b = 4, c = 0$; i.e. $y = 4x - x^2$
The curve has the same basic shape but inverted and displaced, giving roots $x = 0$ and $x = -4$.

11) $a = 1, b = -4, c = 3$; i.e. $y = x^2 - 4x + 3$
Typical parabolic curve, having one turning point and giving two real roots for the quadratic equation $x^2 - 4x + 3 = 0$.

12) $a = 1, b = -4, c = 4$; i.e. $y = (x - 2)^2$
The curve merely touches the axis at the single value $x = 2$.

We conclude that for $y = ax^2 + bx + c$,

i) the sign of a determines whether the curve has a maximum or a minimum
ii) the magnitude of a affects the steepness of the curve;
iii) the ratio of b to a gives the relative displacement sideways, and if $b = 0$ the curve is symmetrical about the y-axis;
iv) the constant c is the intercept on the y-axis;
v) the curve cuts the x-axis at two points if $b^2 > 4ac$ but just touches it if $b^2 = 4ac$ (see section B3.5).

B7.3 Finding the equation of a quadratic
To find the equation of a given parabolic curve we can use the general form $y = ax^2 + bx + c$ and substitute the co-ordinates of any three points on the curve to determine the values of the constants a, b, c.

Example 1 Find the equation of the parabolic curve in fig. B7.4.

We note that the curve cuts the x-axis at the points $(1, 0)$ and $(3, 0)$ and it has a maximum at $(2, 2)$.

Substituting in $y = ax^2 + bx + c$ gives

$$0 = a + b + c \qquad \text{(i}$$
$$0 = 9a + 3b + c \qquad \text{(ii}$$
$$2 = 4a + 2b + c \qquad \text{(iii}$$

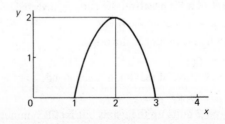

Fig. B7.4

60

Subtracting equation (i) from equation (ii),

$$0 = 8a + 2b \qquad \therefore \quad b = -4a$$

Subtracting equation (i) from equation (iii),

$$2 = 3a + b \qquad \therefore \quad 2 = 3a - 4a \quad \text{and} \quad a = -2$$
$$b = 8$$
$$c = -6$$

$$\therefore \quad y = -2x^2 + 8x - 6$$

Alternatively, we can use the method in section B3.4 for finding the equation of a quadratic with known roots $x = p$ and $x = q$. From this we get an alternative general form for the equation of a quadratic:

$$y = k(x-p)(x-q)$$

Example 2 Using the same parabolic curve, fig. B7.4, we see that the curve cuts the x-axis when $x = 1$ and when $x = 3$. These are the values of x corresponding to p and q, so the equation is

$$y = k(x-1)(x-3)$$

To find the value of the constant k we need the co-ordinates of one other point on the curve, and the point $(2,2)$ is the most convenient.

Substituting the values $x = 2, y = 2$ in $y = k(x-1)(x-3)$ gives $2 = -k$; hence the equation of the curve is

$$y = -2(x-1)(x-3)$$

i.e. $y = -2x^2 + 8x - 6$

B7.4 Graphical solution of quadratic equations

We have already seen that the graph of a quadratic equation is parabolic. For the general equation $y = ax^2 + bx + c$, the curve will have a minimum if a is positive, or a maximum if a is negative. The branches of the curve either side of this turning point are always symmetrical. The curve will cross the x-axis in two places if b^2 is greater than $4ac$, and these will give the roots of the equation, i.e. the required solutions. If b^2 and $4ac$ are equal, the equation is a perfect square and the curve will then just touch the x-axis at its minimum or maximum value. If b^2 is less than $4ac$ then the curve will never cross the x-axis and we say that there are no real roots. If c is zero then the curve will pass through the origin.

Example 1 Plot the curve $y = 2x^2 - 3x - 6$ and hence solve the equations

$$2x^2 - 3x - 6 = 0$$
$$2x^2 - 3x - 5 = 0$$

and $2x^2 - 3x - 9 = 0$

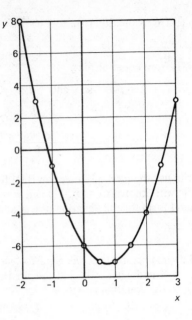

Fig. B7.5 Graph of $y = 2x^2 - 3x - 6$

The graph of $y = 2x^2 - 3x - 6$ is shown in fig. B7.5 and is parabolic with a minimum at $x = \frac{3}{4}$, $y = -7\frac{1}{8}$. It crosses the x-axis where $x = -1.14$ and $x = 2.64$, and thus the solutions of the equation $2x^2 - 3x - 6 = 0$ are $x = 2.64$ or -1.14.

Now the equation $2x^2 - 3x - 5 = 0$ can be put as $2x^2 - 3x - 6 = -1$, and so we can obtain the solutions to this equation simply by reading off on the graph of $y = 2x^2 - 3x - 6$ the values of x for which $y = -1$; i.e. $x = -1$ or $2\frac{1}{2}$.

Similarly, $2x^2 - 3x - 9 = 0$ can be put as $2x^2 - 3x - 6 = 3$, so we read off on the graph the values of x when $y = 3$; i.e. $x = -1.5$ or 3.

Example 2 Draw the graph of $y = x^2$ and use it to solve the equations

$$x^2 = x + 3$$

and $2x^2 - x - 3 = 0$

The graph of $y = x^2$ is shown in fig. B7.6. It is parabolic with a minimum at the origin and is symmetrical about the y-axis.

To use this curve to solve the equation $x^2 = x + 3$, we need to find the values for which x^2 and $x + 3$ are the same. We therefore plot the straight-line graph $y = x + 3$ and find where it crosses the graph of $y = x^2$. At these points, y for each graph is the same; therefore $x^2 = y = x + 3$ at these point and the x-co-ordinates of the points of intersection are the roots of the quad

62

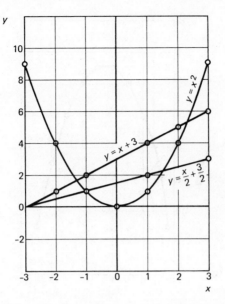

Fig. B7.6

ratic equation $x^2 = x + 3$. The line $y = x + 3$ has been drawn on fig. B7.6, and it crosses the curve where $x = 2.3$ and -1.3.

For the equation $2x^2 - x - 3 = 0$ the same method may be used if the equation is first rearranged into the form $x^2 = \frac{1}{2}x + 1\frac{1}{2}$. The required solutions are then given by the intersections of the straight line $y = \frac{1}{2}x + 1\frac{1}{2}$ with the curve $y = x^2$, which, from fig. B7.6, are $x = 1.5$ and -1.0.

Example 3   Draw the graph of $y = 5x - 2x^2 - 4$ and hence show that the equation $5x - 2x^2 - 4 = 0$ has no real roots. Use the graph to solve the equation $5x - 2x^2 + 2 = 0$. Find also from the graph the value of c in the equation $5x - 2x^2 + c = 0$ for this equation to have equal roots.

The graph of $y = 5x - 2x^2 - 4$ is shown in fig. B7.7. It is parabolic with a maximum, since the coefficient of x^2 is negative. It does not cross the x-axis, so it cannot have any real roots.

The equation $5x - 2x^2 + 2 = 0$ may be put as $5x - 2x^2 - 4 = -6$ and its roots are therefore given by the values of x where $y = -6$ on the graph $5x - 2x^2 - 4 = 0$. From fig. B7.7, the values are $x = 2.85$ and -0.35.

The equation $5x - 2x^2 + c = 0$ will have equal roots if it just touches the x-axis at its maximum. Now the graph of $y = 5x - 2x^2 - 4$ in fig. B7.7 passes below the x-axis at a distance of 0.875 units, so that the equation required is $5x - 2x^2 - 4 = -0.875$; i.e. $5x - 2x^2 - 3.125 = 0$. The value of the constant c for equal roots is thus 3.125.

63

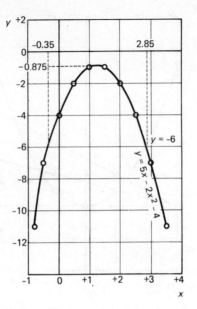

Fig. B7.7

B7.5 Simultaneous equations involving a quadratic

When it is necessary to solve as simultaneous equations one linear equation and one quadratic, this can be done either algebraically or graphically.

Example Solve the simultaneous equations

$$y = 2x^2 - 3x + 2$$

and $y = x + 0.5$.

Algebraically Subtracting to eliminate y,

$$0 = 2x^2 - 4x + 1.5$$

$$0 = (2x - 1)(x - 1.5)$$

$$\therefore \quad x = 0.5 \text{ or } 1.5$$

Graphically The graph of $y = 2x^2 - 3x + 2$ is parabolic and the graph of $y = x + 0.5$ is a straight line which intersects the curve where $x = 0.5$ and 1.5. The graphs are shown in fig. B7.8.

The graphical method usually takes longer than solving the equations algebraically, but it is a useful technique because it can be extended to cope with more complicated equations.

64

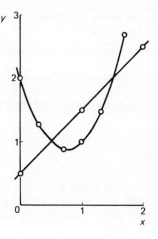

Fig. B7.8

B7.6 Solution of cubics

The cubic equation $x^3 - 1.9x^2 - 3.1x + 4 = 0$ has been solved graphically in three different ways in figs B7.9, B7.10, and B7.11.

1) The first and most obvious way is to plot the graph of the cubic and so determine the points at which it crosses the x-axis. This method is shown in fig. B7.9 and is a perfectly good method provided that neither of the turning points is close to the x-axis.

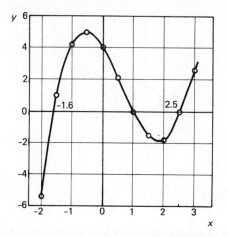

Fig. B7.9 Graph of $y = x^3 - 1.9x^2 - 3.1x + 4$

2) The second method is to split the equation into the familiar cubic $y = x^3$ and a second-degree equation. Thus

$$x^3 - 1.9x^2 - 3.1x + 4 = 0$$

becomes

$$x^3 = 1.9x^2 + 3.1x - 4$$

which shows us that the required solutions are given by the intersection of the two curves

$$y = x^3$$

and $y = 1.9x^2 + 3.1x - 4$

This method is not often very useful because in almost every case it will be found that at least one of the intersections is difficult to judge accurately, as the curves cross one another at a small angle. The greatest accuracy is given by curves which cross at right angles.

This method is used in the graphs of fig. B7.10, and this shows clearly that the negative solution is −1.6, but the solution between +2 and +3 is much harder to see.

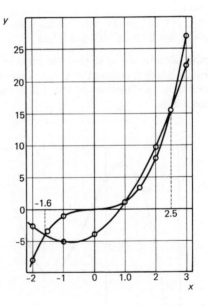

Fig. B7.10 Graphs of $y = x^3$ and $y = 1.9x^2 + 3.1x - 4$

3) Another way of splitting the cubic is to transfer just the term in x and the numerical term so that the graph of this part will be a straight line and therefore simple to plot. Thus, in our example the cubic is split as

$$x^3 - 1.9x^2 = 3.1x - 4$$

and the graphs to be plotted are (fig. B7.11)

$$y = x^3 - 1.9x^2 \quad \text{and} \quad y = 3.1x - 4$$

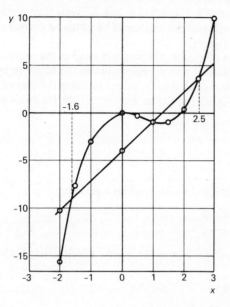

Fig. B7.11 Graphs of $y = x^3 - 1.9x^2$ and $y = 3.1x - 4$

The graph of the cubic $y = x^3 - 1.9x^2 - 3.1x + 4$ shown in fig. B7.9 shows the typical form with a maximum and a minimum clearly discernible. The curve crosses the axis at three points which are the required roots of the cubic equation $x^3 - 1.9x^2 - 3.1x + 4 = 0$.

In fig. B7.10 the two curves are a parabolic curve for $y = 1.9x^2 + 3.1x - 4$ and a simplified cubic $y = x^3$. Here we note that $y = 0$ when $x = 0$, so the curve goes through the origin. By differentiation (see section C1) we could show that the gradient of the curve is zero at the origin, so the curve is horizontal at this point. It can be regarded as a special case in which the maximum and minimum turning points have come together, forming what is known as a *point of inflexion* where the curvature changes.

In fig. B7.11 we have the graph of $y = x^3 - 1.9x^2$. Noting that this factorises to $y = x^2(x - 1.9)$, we see that the curve just touches the axis at the origin and after passing through a minimum it cuts the axis where $x = 1.9$. Dividing the original equation as suggested will always give a turning point at the origin and an intersection with the x-axis easily found by factorisation, so this is a convenient alternative to plotting the more complex original cubic.

67

Closer approximations to roots

The graphical method of obtaining a closer approximation to roots is to plot to an enlarged scale the portion of the graph in the immediate neighbourhood of the approximate solution. In the previous example, a rough graph by any method would reveal the existence of a solution at approximately $2\frac{1}{2}$. If points for x-values 2.3, 2.4, 2.5, 2.6, and 2.7 were then plotted on a much larger scale, the value of the root could be obtained correct to three, if not four, significant figures.

Once one root of a cubic has been obtained to the required degree of accuracy, this solution may be removed from the equation by long division, leaving a quadratic which may be solved by using the formula. It follows that if the remaining quadratic has no real roots, then the cubic has only one solution.

Exercise B7

1 Solve graphically the simultaneous equations

$$y = 6.4x - 4.2$$
$$y = 3.8x + 2.3$$

2 Find graphically the values of v and t which satisfy both of the following equations:

$$v = 95 - 4.9t$$
$$v = 3.5t + 32$$

3 Solve the following pairs of simultaneous equations using the method of substitution or elimination.

(a) $2x = 1.8y - 5.6$ (b) $8.5v + 2.1t - 3.1 = 0$

     $x = 4.1 - 0.3y$ $3.2v - t - 5.2 = 0$

4 Sketch the shapes of the following curves:

(a) $y = 0.4x^2$ (b) $y = 1 + x^2$ (c) $y = (1 + x)^2$

(d) $y = x(1 + x)$ (e) $y = (x - 2)(x - 5)$ (f) $y = 7x - x^2 - 10$

5 Taking values of x from 0 to 5, plot the graph of $y = 2x^2 - 10.7x + 12$ and hence solve the equations

$$2x^2 - 10.7x + 12 = 0$$
$$2x^2 - 10.7x + 14 = 0$$
$$2x^2 - 10.7x + 6 = 0$$

6 Plot the graph of $y = 4x - x^2 + 2$ for x-values -2 to $+5$ and use it to solve the equations $4x - x^2 + 2 = 0$ and $4x = x^2 + 1$. For what value of c would the equation $4x - x^2 + c = 0$ have equal roots?

7 Plot the graph of $y = x^2 - 3x + 1$, taking values of x from 0 to 4. Use the graph to solve the equation $2x^2 - 6x + 1 = 0$. By superimposing a suitable straight-line graph, solve the equation $x^2 - 4x + 2 = 0$.

8 Plot the graph of $y = x^2$ for values of x from -4 to $+4$ and use it to solve the following equations:

a) $x^2 - x - 1 = 0$
b) $x^2 = 2(1-x)$

9 Solve the simultaneous equations

$$y = 3x^2 - 8x + 4$$
$$y = 3 - 2x$$

by plotting graphs over the range from $x = -1$ to $x = +2$.

10 Use graphical methods to find the roots of the following equations:

a) $x^3 - 4.5x^2 - 20x + 28 = 0$
b) $x^3 - 8.2x^2 + 15x + 3.2 = 0$
c) $x^3 - 5.10x^2 - 11.32x + 19.20 = 0$

11 Plot the graph of the equation $y = x^3 + 0.75x^2 - 5x - 26$ for $-4 \leqslant x \leqslant 4$ and hence find the value of the real root to two decimal places.

B8 Exponential graphs and logarithmic scales

B8.1 Exponential growth and decay

Exponential functions have the form $y = a\,e^{kx}$. Such functions are particularly important because natural laws of growth and decay have this form and similar curves can be drawn to illustrate either the growth of bacteria (with unlimited nutrient) or the growth of an investment with compound interest. Likewise, the atomic decay of an initial quantity of a radioactive substance follows a similar pattern to the decay of an electrical charge.

Example 1 The mass of a particular radioactive substance is halved every ten days (i.e. the half-life is ten days). Plot a graph showing the residual amounts of an initial mass of m milligrams over a period of three months.

We can draw up a table of values as follows:

Time (days)	0	10	20	30	40	50	60	70	80	90
Mass (mg)	m	$\dfrac{m}{2}$	$\dfrac{m}{4}$	$\dfrac{m}{8}$	$\dfrac{m}{16}$	$\dfrac{m}{32}$	$\dfrac{m}{64}$	$\dfrac{m}{128}$	$\dfrac{m}{256}$	$\dfrac{m}{512}$

These are shown plotted in fig. B8.1.

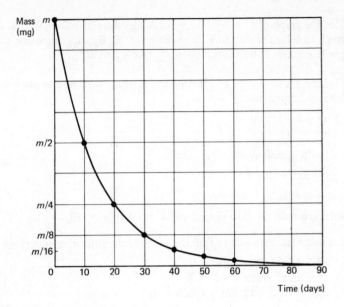

Fig. B8.1 Graph of radioactive decay

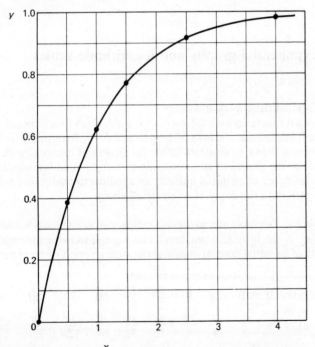

Fig. B8.2 Graph of $y = 1 - e^{-x}$

There are some situations which occur quite naturally where an initial rapid increase is followed by a more gradual growth approaching an ultimate limit. A typical growth curve of this type is illustrated by the graph of $y = 1 - e^{-x}$ shown in fig. B8.2.

Example 2   A $100\,\mu\text{F}$ capacitor is connected in series with a 0.5 MΩ resistor to a 12 V d.c. supply. The voltage v across the capacitor at time t seconds is given by $v = 12(1 - e^{-t/RC})$. Plot a graph of v against t between $t = 0\,\text{s}$ and $t = 180\,\text{s}$.

The graph in fig. B8.3 is plotted using the following table of values

t (s)	0	20	40	60	80	100	120	140	160	180
v (V)	0	3.96	6.61	8.39	9.58	10.38	10.91	11.27	11.51	11.67

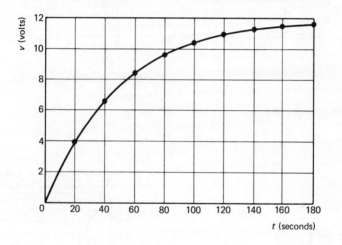

Fig. B8.3 Charging a capacitor

B8.2 Logarithmic scales

Graph paper on which one or both axes have logarithmic scales may be used to verify laws involving exponential or power relationships. A logarithmic scale has graduations at progressively decreasing intervals, like the scales on a slide rule:

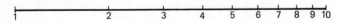

The range of values from 1 to 10 can be covered by a single such scale, or cycle, but values from 1 to 100 would require two cycles one after the other (compare the C and B scales on a slide rule).

71

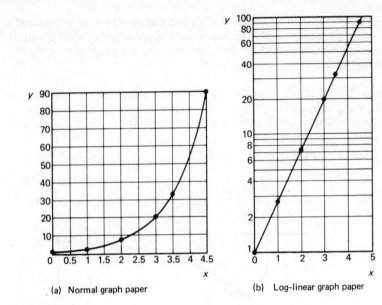

(a) Normal graph paper

(b) Log-linear graph paper

Fig. B8.4 Graph of $y = e^x$

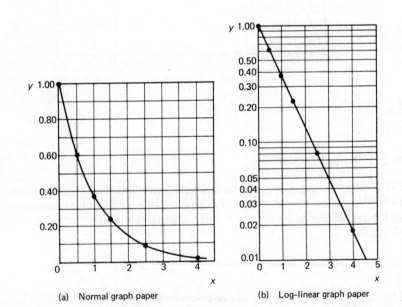

(a) Normal graph paper

(b) Log-linear graph paper

Fig. B8.5 Graph of $y = e^{-x}$

In fig. B8.4(a) and (b), the following x- and y-values for $y = e^x$ are shown plotted (a) on normal graph paper and (b) on graph paper which has a two-cycle logarithmic scale on the vertical axis only. The use of log–linear paper in (b) has reduced the exponential curve to a straight line.

x	0	1.0	2.0	3.0	3.5	4.5
y	1.0	2.7	7.4	20	33	90

Plotting the equivalent graphs for $y = e^{-x}$ (figs B8.5(a) and (b)) gives a similar curve and straight line in the reverse direction. Note that we still use two-cycle log–linear graph paper to reduce the curve to a straight line, but, instead of the logarithmic scale going from 1 to 100, in this case it goes from $\frac{1}{100}$ to 1.

x	0	0.5	1.0	1.5	2.5	4.0
y	1.00	0.61	0.37	0.22	0.08	0.02

We can use the straight line obtained when we plot experimental results from a relationship of the form $y = ae^{kx}$ on log–linear graph paper to determine the constants a and k.

In example 4 of section B6 we reduced $F = Te^{-\mu\theta}$ to a straight-line form to determine the values of the constants T and μ. Figure B8.6 shows an alternative way of plotting the set of values.

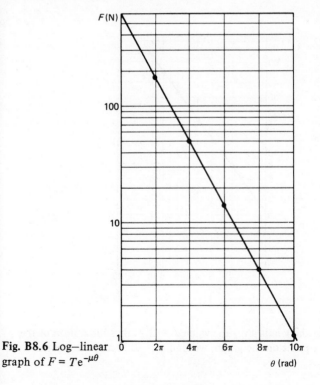

Fig. B8.6 Log–linear graph of $F = Te^{-\mu\theta}$

73

θ (radians)	2π	4π	6π	8π	10π
F (N)	170	48.6	13.8	3.94	1.12

It should be noted that the intercept on the F-axis gives the value of T directly (which in this case can be seen to be approximately 602). The intercept on the θ axis is approximately 10.2π and corresponds to the value of θ for which $F = 1$.

From $F = Te^{-\mu\theta}$, taking logs,

$$\ln F = \ln T - \mu\theta$$

Substituting $F = 1$, $\ln F = 0$; $T = 602$, $\ln T = 6.4$; $\theta = 10.2\pi$ gives

$$\mu = \frac{6.4}{10.2\pi} = 0.20$$

and the law is shown to be $F = 602\,e^{-0.2\theta}$.

For laws of the form $y = ax^n$, taking logs gives $\log y = \log a + n \log x$ but instead of plotting $\log y$ against $\log x$ (as in section B6) we can use log–log paper, which is graph paper with logarithmic scales on both axes, and plot on this the original x- and y-values, as shown in fig. B8.7.

x	2.00	4.00	6.00	8.00	10.00
y	7.88	19.40	32.87	47.77	63.85

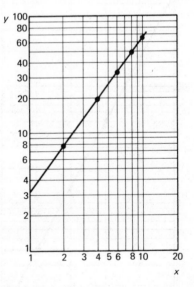

Fig. B8.7

The intercept on the y-axis gives the value $a = 3.2$. By substitution of the co-ordinates of a point on the graph, we find $n = 1.30$ and the law is $y = 3.2x^{1.3}$.

74

Exercise B8

1 The voltage across a capacitor as it discharges is given by $v = E\,e^{-t/RC}$. Plot a graph of v against t between $t = 0\,\text{s}$ and $t = 2\,\text{s}$ if $E = 5\,\text{V}$, $R = 10\,\text{k}\Omega$, and $C = 50\,\mu\text{F}$.

2 The following pairs of values are believed to be connected by a relationship of the form $y = a\,e^{bx}$. Plot a graph on log–linear paper to verify this and estimate the values of the constants a and b.

x	0.20	0.50	0.80	1.20	1.60	2.00
y	0.915	1.31	1.88	3.04	4.91	7.94

3 The table below shows atmospheric pressure p at altitude h. Plot a straight-line graph to verify the exponential relationship $p = a\,e^{bh}$ and find the values of the constants a and b.

h (metres)	0	2000	4000	6000	8000
p (mm)	760	667	586	515	452

4 From the formula $i = E/R\,(1 - e^{-Rt/L})$ for the growth of a current i in time t, given that $E = 300$ when $R = 30$ and $L = 5$, construct a table of values showing values of i when t is 0.01, 0.02, 0.05, 0.10, 0.20, 0.50, 1.00. Plot the curve of current against time and find the value of t when $i = 5.00$.

5 The following table shows values of x and y connected by the relationship $y = ax^n$. Plot these pairs of values on log–log paper and find the values of the constants a and n.

x	0.6	0.9	1.5	2.4	3.2
y	8.94	18.9	48.7	116	198

6 The following measurements of pressure and volume of a gas were taken under conditions of adiabatic expansion. Verify that these results correspond to the law $pV^k = C$ and find the values of the constants k and C.

V	5	6	8	10	12	15
p	5.376	4.157	2.771	2.023	1.564	1.142

B9 Polar co-ordinates and polar graphs

B9.1 Co-ordinates r and θ

It is sometimes more useful to identify a point such as P in fig. B9.1 by the *polar* co-ordinates (r, θ) rather than the rectangular or *Cartesian* co-ordinates (x, y). When this method is used, r is the distance from the origin or *pole* O to the point P, and θ is the angle (measured in degrees or radians) which the line OP makes with the x-axis.

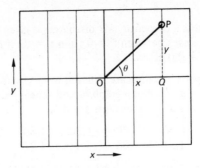

Fig. B9.1 Polar and rectangular co-ordinates of a point

The angle θ is positive when measured anticlockwise, and negative when measured clockwise from the x-axis. This means that $(2, 330°)$ and $(2, -30°)$ are the same point.

B9.2 Evaluation of r and θ

We can find the value of r by applying the theorem of Pythagoras to triangle OPQ in fig. B9.1:

$$r^2 = x^2 + y^2$$

$$\therefore \quad r = \sqrt{(x^2 + y^2)}$$

From the same triangle,

$$\tan \theta = y/x$$

$$\therefore \quad \theta = \arctan(y/x)$$

B9.3 Changing co-ordinates

To convert from Cartesian to polar co-ordinates, we can make use of the expressions for r and θ given in section B9.2. However, if x or y has a negative value (i.e. the point lies in the second, third, or fourth *quadrant*) we must be careful to arrive at the correct value for θ. Your calculator will not give the right answer automatically. In these cases it is better to leave out the minus signs during calculations, and to use a diagram in deciding which angle we require.

Consider the point P at $(-4.5, 6.3)$ in fig. B9.2(a). Finding r does not present a problem for, although the value of x is negative in this case, values of x^2 and y^2 are always positive and thus

$$r = \sqrt{(x^2 + y^2)} = \sqrt{(20.25 + 36.69)} = 7.74$$

If we draw the triangle QPO (see fig. B9.2(b)), including the lengths we know, then

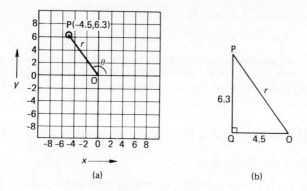

Fig. B9.2

$$\angle O = \arctan\frac{6.3}{4.5} = 54.46°$$

and $\qquad \theta = 180° - \angle O$

$\therefore \qquad \theta = 125.54°$

The polar co-ordinates of P are therefore $(7.74, 125.54°)$.

Conversely, to find the Cartesian co-ordinates of a point given its polar co-ordinates, we can see from triangle OPQ in fig. B9.1 that

$$\frac{x}{r} = \cos\theta \quad \text{or} \quad x = r\cos\theta$$

and $\dfrac{y}{r} = \sin\theta \quad \text{or} \quad y = r\sin\theta$

These relationships can be used to evaluate x and y. Again, care should be taken when dealing with points in the second, third, and fourth quadrants. If you are using a calculator to obtain values of $\cos\theta$ and $\sin\theta$, it will give a + or − sign automatically for any angle θ. In the case of point P in fig. B9.3(a),

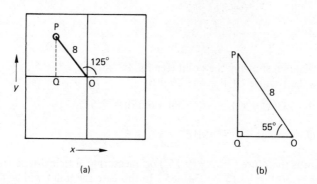

Fig. B9.3

77

$$x = r \cos \theta = 8 \cos 125° = 8\,(-0.5736) = -4.59$$

and $\quad y = r \sin \theta = 8 \sin 125° = 8\,(0.8192) \quad = 6.55$

To obtain these same values using tables, we first sketch the triangle QPO (fig. B9.3(b)), where $\angle O = 180° - 125° = 55°$. The x-co-ordinate of P is the length OQ and is negative. The y-co-ordinate is PQ and is positive. From the triangle,

$$\frac{OQ}{8} = \cos 55° = 0.5736 \quad \text{and} \quad \frac{PQ}{8} = \sin 55° = 0.8192$$

$\therefore \quad$ OQ $= 8 \times 736 = 4.59 \qquad \therefore$ PQ $= 8 \times 0.8192 = 6.55$

Now, putting in the correct signs, we can say the point P is at $(-4.59, 6.55)$.

Example 1 Find the polar co-ordinates of the point $(5, 20)$.

$$r = \sqrt{(x^2 + y^2)} = \sqrt{(25 + 400)} = 20.62$$
$$\theta = \arctan(y/x) = \arctan(20/5) = 75.96°$$

i.e. the polar co-ordinates are $(20.62, 75.96°)$.

Example 2 What are the polar co-ordinates of the point $(7.2, -2.5)$?

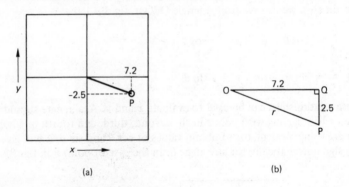

Fig. B9.4

Draw a diagram showing the point (see fig. B9.4(a)).
    From the triangle OQP in fig. B9.4(b),

$$r = \sqrt{(x^2 + y^2)} = \sqrt{(51.84 + 6.25)} = 7.62$$

and $\quad \angle O = \arctan \dfrac{2.5}{7.2} = 19.15°$

The point is therefore $(7.62, -19.15°)$ or, measuring anticlockwise from the x-axis, $\theta = 360° - 19.15° = 340.85°$. In this case the point will be $(7.62, 340.85°)$.

78

Example 3   What are the Cartesian co-ordinates of a point with polar co-ordinates $(12, 28°)$?

$$x = r \cos \theta = 12 \cos 28° = 12 \times 0.8829 = 10.60$$
$$y = r \sin \theta = 12 \sin 28° = 12 \times 0.4695 = 5.63$$

i.e. the point is $(10.60, 5.63)$.

Example 4   Find the Cartesian co-ordinates of A in fig. B9.5(a).

Method 1   With your calculator, use the definitions from section B9.3 to find x and y directly:

$$x = r \cos \theta = 2.8 \cos 217.5° = 2.8 \,(-0.7934) = -2.22$$
$$y = r \sin \theta = 2.8 \sin 217.5° = 2.8 \,(-0.6088) = -1.70$$

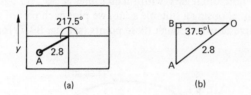

(a)   (b)

Fig. B9.5

Method 2   Draw the triangle ABO shown in fig. B9.5(b), in which $LO = 217.5° - 180° = 37.5°$. From this, using tables or your calculator,

$$\frac{OB}{2.8} = \cos 37.5° \qquad \text{and} \qquad \frac{AB}{2.8} = \sin 37.5°$$

$$\therefore \quad OB = 2.8 \times 0.7934 \qquad \therefore \quad AB = 2.8 \times 0.6088$$
$$= 2.22 \qquad\qquad\qquad = 1.70$$

Note from the diagram that both x and y co-ordinates are negative (see fig. B9.5(a)). The point A is therefore $(-2.22, -1.70)$.

B9.4 Plotting graphs using polar co-ordinates
To draw a polar graph we need to know values of r for a corresponding range of θ values (in degrees or radians). To plot the point $(5, 30°)$ on a graph, first draw a horizontal axis and then a line at $30°$ to it, measured anticlockwise, using a protractor. Now measure 5 units along this line from the origin or *pole* O, and mark the point (fig. B9.6(a)).

For the point $(-4, 45°)$, the value given for r is a negative number. We still draw a line at $45°$ to the horizontal axis, but, to plot the point on the graph, the measurement of 4 units is made on the other side of the pole, as shown in fig. B9.6(b).

79

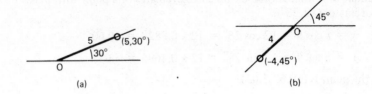

Fig. B9.6

Example Plot a polar graph using the following table, which gives values of r for $\theta = 0°$ to $180°$ at $30°$ intervals, obtained from the equation $r = 1 + \cos \theta$

θ	$0°$	$30°$	$60°$	$90°$	$120°$	$150°$	$180°$
r	2.00	1.87	1.50	1.00	0.50	0.13	0

We first draw a horizontal axis, with further lines at $30°$, $60°$, $90°$, $120°$, and $150°$ to it. Then, choosing a suitable scale, we plot each pair of co-ordinates in turn. The curve drawn through these points (see fig. B9.7) forms one half of a *cardioid*.

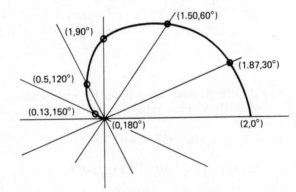

Fig. B9.7 A graph drawn using polar co-ordinates

The spiral
Curves with equations of the form $r = A\theta$ (where A is a constant and θ is in radians) are called *spirals*. Figure B9.8 shows $r = \theta$ drawn between $\theta = 0$ and 4π rad.

When drawing polar graphs it is not necessary to use squared paper. Polar graph paper can be obtained for special purposes.

Exercise B9
1 Convert to polar co-ordinates (a) (4.2, 6.5); (b) (−8, 14); (c) (−0.12, −0.35); (d) (125, −247).

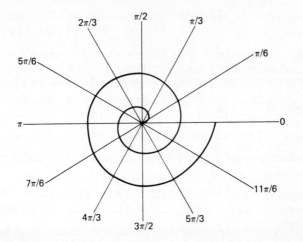

Fig. B9.8 The spiral $r = \theta$

2   Convert to Cartesian co-ordinates (a) $(9, 78°)$; (b) $(1.25, 165°30')$;
(c) $(32.5, 7\pi/6)$; (d) $(0.77, -27°)$; (e) $(-15, 111°)$.

3   A man starts walking in a straight line in a direction 20° north of west. After 3 kilometres he turns a further 10° northward and walks another 4 km. What are his new co-ordinates west and north of his starting point?

4   A numerically controlled machine-tool is to drill nine holes in the circular cover-plate shown in fig. B9.9. If hole A is drilled first, what movements in the x- and y-directions must the drill then make to be in position for hole B?

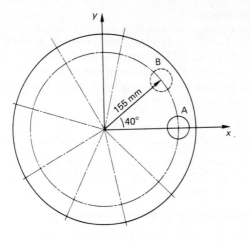

Fig. B9.9

5   Show the points $(7.5, -55°)$ and $(-5, 240°)$ on the same polar graph.

6   Use the following table of values to plot one quadrant of the polar graph for $r - 10 \sin 2\theta$.

θ (degrees)	0	10	20	30	40	50	60	70	80	90
r (mm)	0	34	64	87	98	98	87	64	34	0

7 The table of values below gives measurements of the radius r of a cam (used in a carburetter choke mechanism) against angle θ at $15°$ intervals. Draw an outline of the cam, twice full-size, by plotting the points on a polar graph.

θ (radians)	0	$\frac{\pi}{12}$	$\frac{\pi}{6}$	$\frac{\pi}{4}$	$\frac{\pi}{3}$	$\frac{5\pi}{12}$	$\frac{\pi}{2}$	$\frac{7\pi}{12}$	$\frac{2\pi}{3}$	$\frac{3\pi}{4}$	$\frac{5\pi}{6}$	$\frac{11\pi}{12}$	π
r (mm)	10	12	14	16	18	20	22	24	26	28	30	32	34

8 The following values of G and θ were obtained from a stability test on a servomechanism:

θ (degrees)	150	160	170	180	190	200	210	220
G (mm)	14.5	22.0	29.5	37.5	45.5	55.0	66.5	81.0

Draw a full-size polar graph using these points, and measure the polar co-ordinates of the point where your line crosses a 50 mm radius circle, centred on the pole.

9 Draw the spiral $r = 2\theta$ between $\theta = 0$ and 3π rad, using a scale of 1 unit to 4 mm and plotting points every $\pi/4$ radians.

B10 Boolean algebra

B10.1 Two-state devices
Digital electronic circuits, which now have applications in almost every branch of technology, are based on a *two-state* system of voltage levels. A simple switch can be used as a model to help us to understand some of the theory involved in the design and simplification of these circuits.

The switch in fig. B10.1(a) is open, while in fig. B10.1(b) an input A has energised the relay coil and closed the switch.

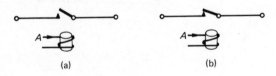

(a)   (b)

Fig. B10.1 A simple switch is a two-state device

B10.2 The binary system of numbers
Most of us are familiar with the *binary system* of numbers, which uses only 0 and 1 instead of all the figures from 0 to 9 used in the decimal system.

Binary arithmetic is most useful when dealing with two-state systems. Referring to the switch in fig. B10.1, we can relate the two states of input A to the binary digits 0 and 1. We shall let $A = 0$ represent the input when the switch is open and let $A = 1$ be the input when the switch is closed.

B10.3 The AND, OR, and NOT functions

AND, OR, and NOT are names we give to the basic functions carried out by switching elements or *gates* in digital electronic circuits. We can demonstrate their meaning using the simple switch as a model.

The AND function

Consider the arrangement of two switches in fig. B10.2(a), where they are connected in *series*. When $A = 1$ the first switch will close, and when $B = 1$ the second switch will close; but the lamp will light only when both $A = 1$ *and* $B = 1$.

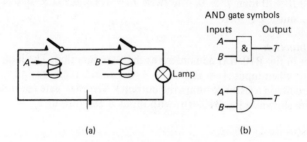

Fig. B10.2 The AND function and gate symbols

We express the AND function in *Boolean algebra* as $A.B$, using a *dot* to signify A AND B. (Note that Boolean algebra is used in the design of two-state systems to analyse and simplify circuits. We shall consider further rules of this algebra in later sections.)

The symbols most frequently used to represent the AND gate are shown in fig. B10.2(b). Two inputs, A and B, are shown although gates are also manufactured with a larger number of inputs. The output T will be 1 only when both A and B are 1; any other combination of inputs will give 0 at the output. We write $A.B = T$ for the gate, and thus for the four combinations of input we have $0.0 = 0$, $0.1 = 0$, $1.0 = 0$, and $1.1 = 1$.

The OR function

To represent the OR function, we connect two switches in parallel (see B10.3(a)).

If either of the switches is closed the lamp will light, and obviously the lamp will also light if both switches are closed. The OR gate symbols are shown in fig. B10.3(b) with inputs A and B and output T. In Boolean algebra we use a *plus sign* to signify the OR function, so that for the OR gate $A + B = T$. There are again four combinations of A and B to consider: if

83

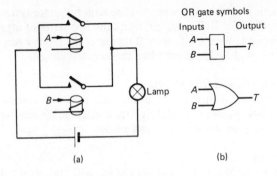

Fig. B10.3 The OR function and gate symbols

$A = 0$ and $B = 0$ then $T = 0$, otherwise $T = 1$; i.e. $0 + 0 = 0$, $0 + 1 = 1$, $1 + 0 = 1$, and $1 + 1 = 1$.

The NOT function

The circuit in fig. B10.4(a) demonstrates the NOT function. The lamp is normally lit when input $A = 0$. If $A = 1$, the switch is closed and the lamp is short-circuited (resistor R limits the current). The NOT gate (or invertor) symbols are shown in fig. B10.4(b) with input A and output T.

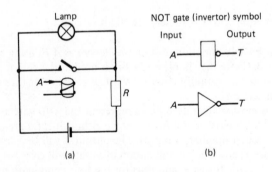

Fig. B10.4 The NOT function and gate symbols

When $A = 0$, $T = 1$; and when $A = 1$, $T = 0$. We write $T = \bar{A}$ with a *bar* over the A to define the NOT function in Boolean algebra. $\bar{A}$ (we say 'NOT-A') is the opposite or inverse of A, so that $\bar{0} = 1$ and $\bar{1} = 0$.

B10.4 Simple combinations of gates

Using gates connected together, it is possible to design circuits which will have a particular series of outputs for a given combination of input states.

84

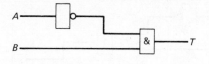

Fig. B10.5

Consider the circuit in fig. B10.5. Inputs A and B can both be either 0 or 1, and there are therefore four possible combinations of inputs (i.e. 00, 01, 10, and 11). The output T will be 1 only when both inputs to the AND gate are 1. The only way this will happen is if $B = 1$ and $A = 0$, since the NOT-gate output will be $\bar{A}$ at all times. We can write the equation $T = \bar{A}.B$ to describe the output.

Example 1   What is the output T when $A = 1, B = 1$ and $C = 0$ in the circuit in fig. B10.6?

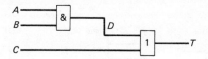

Fig. B10.6

If we consider the point D at the output of the AND gate,

$$D = A.B$$

and, with $A = 1$ and $B = 1$,

$$D = 1.1 = 1$$

The inputs to the OR gate are D and C, so that

$$T = D + C$$

With $C = 0$,

$$T = 1 + 0 = 1$$

i.e., when $A = 1, B = 1$, and $C = 0$, the output T is 1.

Example 2   Determine the output T for the combination of gates in fig. B10.7 when $A = 1, B = 0$, and $C = 1$.

First we label the signals D and E on the circuit and then work from input to output as follows:

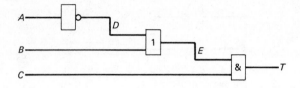

Fig. B10.7

$$D = \overline{A} = \overline{1} = 0$$
$$E = D + B = 0 + 0 = 0$$
$$T = E.C = 0.1 = 0$$

i.e., when $A = 1$, $B = 0$, and $C = 1$, the output T is 0.

B10.5 Truth tables

For all possible input conditions, we can write down the outputs of a single gate or a complete circuit in a *truth table*. In fig. B10.8, truth tables have been drawn for the AND, OR, and NOT gates. The truth table for a combination

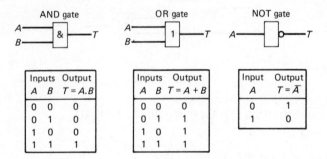

Fig. B10.8 Truth tables for AND, OR, and NOT gates

of gates is drawn in the same way, with all possible inputs in columns on the left and output states corresponding to these inputs on the right. In fig. B10.9 we have an AND gate with a NOT gate connected to one input.

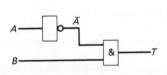

Inputs		Output	
A	B	$\overline{A}$	$T = \overline{A}.B$
0	0	1	0
0	1	1	1
1	0	0	0
1	1	0	0

Fig. B10.9 Truth table for a combination of gates

In this example it is useful to include in the truth table a column for $\bar{A}$ in addition to the inputs A and B. Since $T = \bar{A}.B$, the output states for each line of the table can be obtained by referring to columns $\bar{A}$ and B. Adding *intermediate functions* to the table in this way helps us to avoid errors which may occur if we attempt to write down output states directly from the inputs for circuits with a large number of gates.

Example Draw a truth table for the combination of gates in fig. B10.10.

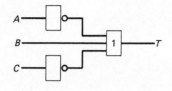

Fig. B10.10

Columns for $\bar{A}$ and $\bar{C}$ should be added to the truth table for this circuit, since $T = \bar{A} + B + \bar{C}$. The complete table has eight lines of possible input states, since the circuit has three inputs. In general, if a circuit has n inputs there will be 2^n lines in its truth table.

Inputs A B C			$\bar{A}$ B $\bar{C}$			Output $T = \bar{A} + B + \bar{C}$
0	0	0	1	0	1	1
0	0	1	1	0	0	1
0	1	0	1	1	1	1
0	1	1	1	1	0	1
1	0	0	0	0	1	1
1	0	1	0	0	0	0
1	1	0	0	1	1	1
1	1	1	0	1	0	1

Note that the column for B states has been included a second time in the table, between $\bar{A}$ and $\bar{C}$. This makes it easier to obtain the states for T and reduces the chances of misreading a 1 or a 0 across the table. We can see from the truth table that output T is 1 except when $A = 1, B = 0$, and $C = 1$.

B10.6 The commutative, associative, and distributive laws
Boolean algebra is used by the designer to analyse and simplify digital systems so that the most suitable choice of gates can be made. We shall now study some of the laws of Boolean algebra.

87

The commutative law

The commutative law states that we can change the order of variables in an expression containing only AND functions or OR functions without changing the meaning of the expression; thus

$$A + B = B + A$$

$$A.B = B.A$$

In practical terms this implies that it is irrelevant which inputs we use on a particular OR gate or AND gate. The following truth tables demonstrate this law.

A	B	$A + B$	$B + A$
0	0	0	0
0	1	1	1
1	0	1	1
1	1	1	1

A	B	$A.B$	$B.A$
0	0	0	0
0	1	0	0
1	0	0	0
1	1	1	1

The truth-table columns for $A + B$ and $B + A$ are exactly the same, as are those for $A.B$ and $B.A$ for all combinations of A and B. When columns of a truth table are the same for two expressions in this way, we say that the expressions are *equivalent*. We shall now use this technique to demonstrate other laws of Boolean algebra.

The associative law

This law states that we can perform identical functions in any order; thus

$$A + B + C = (A + B) + C = A + (B + C)$$

$$A.B.C = (A.B).C = A.(B.C)$$

We use brackets as in ordinary arithmetic to group operations together; so $(A + B) + C$ means that we first carry out the OR operation on A and B and then use the result of this in a second OR operation with C. The following truth table demonstrates the law for the OR function.

A	B	C	$(A + B)$	$(B + C)$	$(A + B) + C$	$A + (B + C)$
0	0	0	0	0	0	0
0	0	1	0	1	1	1
0	1	0	1	1	1	1
0	1	1	1	1	1	1
1	0	0	1	0	1	1
1	0	1	1	1	1	1
1	1	0	1	1	1	1
1	1	1	1	1	1	1

Since the two columns on the right of the table are the same, the expressions $(A + B) + C$ and $A + (B + C)$ are equivalent. By drawing a truth table, you may like to show that $(A.B).C$ and $A.(B.C)$ are also equivalent.

The distributive law
This law is expressed in two forms:

$$A.(B + C) = A.B + A.C$$

$$A + (B.C) = (A + B).(A + C)$$

Again we can use a truth table to show that the expressions on each side of these equations are equivalent. Taking the first equation as an example,

A	B	C	B + C	A.B	A.C	A.(B+C)	A.B + A.C
0	0	0	0	0	0	0	0
0	0	1	1	0	0	0	0
0	1	0	1	0	0	0	0
0	1	1	1	0	0	0	0
1	0	0	0	0	0	0	0
1	0	1	1	0	1	1	1
1	1	0	1	1	0	1	1
1	1	1	1	1	1	1	1

B10.7 Other useful laws
In addition to the commutative, associative, and distributive laws, the relationships in fig. B10.11 are always true, as is demonstrated by a truth table in each case.

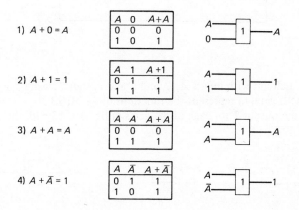

Fig. B10.11 Some useful laws from Boolean algebra (*cont'd on page 90*)

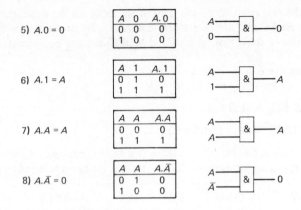

	A	0	A.0
5) A.0 = 0	0	0	0
	1	0	0

	A	1	A.1
6) A.1 = A	0	1	0
	1	1	1

	A	A	A.A
7) A.A = A	0	0	0
	1	1	1

	A	A	A.$\bar{A}$
8) A.$\bar{A}$ = 0	0	1	0
	1	0	0

Fig. B10.11 Some useful laws from Boolean algebra (*cont'd*)

Exercise B10
1 If any of the switches in fig. B10.12 is closed when its input is 1, find whether the lamp is on or off in the following cases: (a) $A = 1, B = 0, C = 0$; (b) $A = 1, B = 1, C = 0$; (c) $A = 0, B = 1, C = 1$; (d) $A = 1, B = 1, C = 1$.

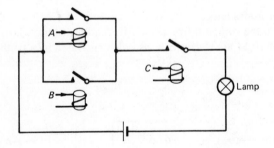

Fig. B10.12

2 Draw a circuit with inputs A, B, and C and output T, using one OR gate and one AND gate, to represent the functions in fig. B10.12.
3 Find the output T for each of the gates in fig. B10.13 with the given inputs.

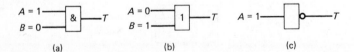

Fig. B10.13

90

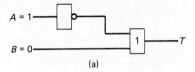

(a)

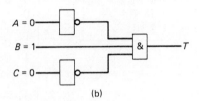

(b)

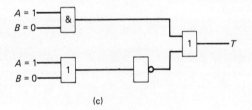

(c)

Fig. B10.14

4 Find the output T for each of the circuits in fig. B10.14 with the inputs shown.

5 Show that the circuit in fig. B10.15 has the same outputs as a two-input OR gate for all inputs A and B except when $A = 1$ and $B = 1$.

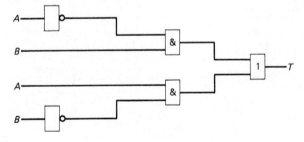

Fig. B10.15

6 Draw truth tables for each of the circuits in fig. B10.14.

7 Draw a truth table for the expression $(A + \overline{B}).\overline{C}$.

8 Complete the following truth tables:

(a)

A	B	$A + \overline{B}$
0	0	
0	1	
1	0	
1	1	

(b)

A	B	$\overline{A}.B$
0	0	
0	1	
1	0	
1	1	

(c)

A	B	C	$A + \overline{B} + C$
0	0	0	
0	0	1	
0	1	0	
0	1	1	
1	0	0	
1	0	1	
1	1	0	
1	1	1	

C   Calculus

C1   Differentiation

C1.1 Average and instantaneous gradients

The graph of $y = 4x - \frac{1}{2}x^2$ is shown in fig. C1.1, plotted from the following set of values:

x	0	1	2	3	4	5	6	7	8
y	0	3.5	6	7.75	8	7.75	6	3.5	0

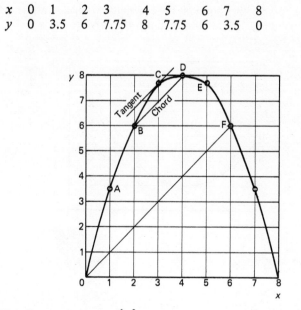

Fig. C1.1 Graph of $y = 4x - \frac{1}{2}x^2$

The curve is parabolic and is symmetrical about a vertical line through D, which is the highest point. Taking a *chord* to be a line joining two points on the curve and a *tangent* to be a line which just touches the curve, fig. C1.1 shows two chords and a tangent, which in this case all happen to be parallel.

We can see quite easily from the diagram or from the table of values that chords BD and OE have the same gradient of 1.0, i.e. they slope at an angle of 45° to the x-axis. It is more difficult to sketch the curve accurately enough to be able to draw tangents where required and estimate their gradients. It is even more difficult to estimate the slope of the actual curve.

93

By means of a short chord such as BD we can determine the average gradient of the section of the curve between points B and D. We can see from the diagram that it is not really valid to use chord OE to estimate the average gradient of the curve between O and E, since the actual gradient of the curve varies from +4 at 0, through zero at D, down to −1 at E. From this it is clear that shorter chords are preferable in estimating average gradients.

The gradient of the curve at any point is equivalent to the gradient of the tangent at that point. The gradient of the tangent can be estimated from the graph, bearing in mind that the gradient of the tangent will be close to the gradient of a short chord across adjacent points. To determine the gradient of the tangent accurately requires the use of calculus.

C1.2 Gradient of chord and tangent

Consider the curve $y = x^2$, a portion of which is shown in fig. C1.2. Let P be the point $(1,1)$. Q is another point on the curve and we shall consider it to start from the point $(5,25)$ and to move down the curve towards P.

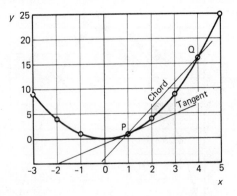

Fig. C1.2 Graph of $y = x^2$

When Q is at $(5,25)$, the gradient of chord PQ is $\dfrac{25-1}{5-1} = \dfrac{24}{4} = 6$

When Q is at $(4,16)$, the gradient of chord PQ is $\dfrac{16-1}{4-1} = \dfrac{15}{3} = 5$

When Q is at $(3,9)$, the gradient of chord PQ is $\dfrac{9-1}{3-1} = \dfrac{8}{2} = 4$

When Q is at $(2,4)$, the gradient of chord PQ is $\dfrac{4-1}{2-1} = \dfrac{3}{1} = 3$

If we now take a point an equivalent distance on the other side of P,

When Q is at $(0,0)$ the gradient of chord PQ is $\dfrac{1-0}{1-0} = 1$

The gradients of the chords thus form a series:

6, 5, 4, 3, −, 1

The missing gradient is that for $x = 1$, but when Q reaches this point it coincides with point P. If the chord PQ is regarded as being part of a longer straight line which simply intersects the curve at the two points P and Q, then, when Q coincides with P, the straight line becomes a tangent to the curve at the point where P and Q coincide. From the series which we have obtained it would therefore seem reasonable to assume that the gradient of the tangent at P would be 2. Let us now look at this a little more closely by considering values for Q close to P.

When Q is at $(1.5, 2.25)$ the gradient of chord PQ is $\dfrac{1.25}{0.5} = 2.5$

When Q is at $(1.4, 1.96)$ the gradient of chord PQ is $\dfrac{1.96}{0.4} = 2.4$

When Q is at $(1.3, 1.69)$ the gradient of chord PQ is $\dfrac{0.69}{0.3} = 2.3$

When Q is at $(1.2, 1.44)$ the gradient of chord PQ is $\dfrac{0.44}{0.2} = 2.2$

When Q is at $(1.1, 1.21)$ the gradient of chord PQ is $\dfrac{0.21}{0.1} = 2.1$

When Q is at $(1.01, 1.0201)$ the gradient of chord PQ is $\dfrac{0.0201}{0.01} = 2.01$

When Q is at $(1.001, 1.002\,001)$ the gradient of chord PQ is $\dfrac{0.002\,001}{0.001} = 2.001$

It thus becomes increasingly obvious that, the closer Q approaches to P, the nearer the gradient is to 2.

C1.3 Differentiation from first principles

Now, instead of selecting some special values, let us take a general case with co-ordinates (x, y) for P, with Q a point very close to it on the curve $y = x^2$, fig. C1.2. Let the x co-ordinate for Q be $x + \delta x$, where δx is a small increase in x. (δ is the Greek letter '*delta*'.)

Since $y = x^2$, letting x increase by a small amount δx results in a corresponding small change δy in y; thus

$$y + \delta y = (x + \delta x)^2$$

or $\quad y + \delta y = x^2 + 2x\delta x + (\delta x)^2$

Subtracting the original, gives

$$\delta y = 2x\delta x + (\delta x)^2$$

Dividing by δx to obtain the gradient of the chord,

$$\frac{\delta y}{\delta x} = 2x + \delta x$$

Now, as P approaches Q, δx becomes smaller until it becomes zero when P and Q coincide. At this point, the gradient of the chord we have been considering has become the gradient of the tangent to the curve at that point. Changes which take place as the initially small δx becomes progressively even smaller are said to take place 'in the limit as δx tends to zero', so we can say that, in the limit as δx tends to zero, the gradient of the chord PQ is equal to the gradient of the tangent at Q. We then no longer use $\delta y/\delta x$ but dy/dx;

i.e. in the limit as δx tends to zero, $\dfrac{\delta y}{\delta x} \to \dfrac{dy}{dx}$

or, in symbols,

$$\lim_{\delta x \to 0}\left(\frac{\delta y}{\delta x}\right) = \frac{dy}{dx}$$

Thus, for the case we are considering, where $y = x^2$,

$$\frac{\delta y}{\delta x} = 2x + \delta x$$

$$\therefore \quad \lim_{\delta x \to 0}\left(\frac{\delta y}{\delta x}\right) = \frac{dy}{dx} = 2x$$

For example, when $x = 1$ (as in section C1.2),

$$\frac{dy}{dx} = \text{gradient of tangent} = 2$$

dy/dx is known as 'the *differential coefficient* of y with respect to x' and the process of determining it is known as *differentiation*. When each step of the process is worked through as above, it is known *as differentiation from first principles*.

Example Differentiate from first principles $y = 1/x$.

Let x increase by a small amount δx.
Let the corresponding increase in y be δy.

Then $\quad y + \delta y = \dfrac{1}{x + \delta x}$

Subtracting the original equation,

$$\delta y = \frac{1}{x + \delta x} - \frac{1}{x} = \frac{x - (x + \delta x)}{x(x + \delta x)} = \frac{-\delta x}{x(x + \delta x)}$$

Dividing by δx,

$$\frac{\delta y}{\delta x} = \frac{-1}{x(x + \delta x)} = \frac{-1}{x^2 + x\delta x}$$

In the limit as $\delta x \to 0$, $\delta y/\delta x \to dy/dx$

$$\therefore \quad \frac{dy}{dx} = -\frac{1}{x^2}$$

C1.4 Differentiation of $y = x^n$

Consider the equation $y = x^n$. Let x increase by a small amount δx, and the corresponding increase in y be δy; then

$$y + \delta y = (x + \delta x)^n$$

If we let $n = 3$, i.e. $y = x^3$,

$$y + \delta y = (x + \delta x)^3$$

Expanding the right-hand side and collecting terms, we get

$$y + \delta y = x^3 + 3x^2\delta x + 3x(\delta x)^2 + (\delta x)^3$$

Subtracting $y = x^3$,

$$\delta y = 3x^2\delta x + 3x(\delta x)^2 + (\delta x)^3$$

and, dividing by δx,

$$\frac{\delta y}{\delta x} = 3x^2 + 3x\delta x + (\delta x)^2$$

In the limit as $\delta x \to 0$, $\delta y/\delta x \to dy/dx$

$$\therefore \quad \frac{dy}{dx} = 3x^2$$

We have already seen in section C1.3 that

when $y = x^2$ $\qquad \dfrac{dy}{dx} = 2x$

when $y = \dfrac{1}{x} = x^{-1}$ $\quad \dfrac{dy}{dx} = -\dfrac{1}{x^2} = -x^{-2}$

and we know that the curve $y = x$ has a gradient 1, i.e.

when $y = x$ $\qquad \dfrac{dy}{dx} = 1 = x^0$

Considering these results in terms of the equation $y = x^n$, we see that

if $n = 3$, i.e. $y = x^3$, $\qquad dy/dx = 3x^2$

if $n = 2$, i.e. $y = x^2$, $\qquad dy/dx = 2x$

97

if $n = 1$, i.e. $y = x$, $dy/dx = x^0$

if $n = -1$, i.e. $y = x^{-1}$, $dy/dx = -x^{-2}$

Comparing these, we can deduce that, in general,

if $\quad y = x^n \qquad \dfrac{dy}{dx} = nx^{n-1}$

This result is true for *all* values of n, including negative and fractional values, and its use as a formula enables us to short-cut the process of differentiation from first principles in many cases.

Examples

a) $y = x^4$ $\qquad\qquad\qquad \dfrac{dy}{dx} = 4x^3$

b) $y = \dfrac{1}{x^3} = x^{-3}$ $\qquad \dfrac{dy}{dx} = -3x^{-4} = -\dfrac{3}{x^4}$

c) $y = \sqrt{x^3} = x^{\frac{3}{2}}$ $\qquad \dfrac{dy}{dx} = \dfrac{3}{2}x^{\frac{1}{2}} = \dfrac{3\sqrt{x}}{2}$

d) $y = \dfrac{1}{\sqrt{x}} = x^{-\frac{1}{2}}$ $\qquad \dfrac{dy}{dx} = -\dfrac{1}{2}x^{-\frac{3}{2}} = \dfrac{-1}{2\sqrt{x^3}}$

e) $y = 1 = x^0$ $\qquad\qquad \dfrac{dy}{dx} = 0$

C1.5 Differentiation of constants

When a power of x is multiplied by a constant, that constant remains unchanged by the process of differentiation; i.e.

if $\quad y = ax^n, \qquad \dfrac{dy}{dx} = anx^{n-1}$

The product *an* will then be combined into a single new constant.

If any isolated constant is differentiated, the result will be zero.

For the straight-line equation $y = mx$, $dy/dx = m$, i.e. the straight line has gradient m.

The equation $y = c$ gives a straight line parallel to the x-axis, i.e. with gradient zero. Thus if $y = c$ then $dy/dx = 0$.

C1.6 Differentiation of a sum of several terms

The rule is to differentiate term by term and combine the results.

Examples

a) $y = ax^2 + bx + c$ $\qquad\qquad\qquad\qquad \dfrac{dy}{dx} = 2ax + b$

b) $y = 4x^3 - 2$ $\dfrac{dy}{dx} = 12x^2$

c) $y = (2x + 1)(x - 3) = 2x^2 - 5x - 3$ $\dfrac{dy}{dx} = 4x - 5$

C1.7 Choice of variable

Before an equation can be differentiated in the way shown above, it is necessary for the equation to be arranged so that there is only one variable on each side of the equals sign. The symbols used need not be x and y, but it is conventional to use letters from the second half of the alphabet to represent variables. In the equation $s = 20t - 4.9t^2$ we have t as the independent variable (in place of x). If this equation had been given as $s = ut + \frac{1}{2}at^2$ it would not have been so obvious which letters represented the variables. Given that u represented an initial velocity of 20 m/s upwards and that a was the acceleration of 9.8 m/s^2 downwards, we could substitute the values of these constants and thus obtain the first equation. Having established which symbols represent the variables, it is then possible to differentiate:

$$s = 20t - 4.9t^2$$

$\therefore \quad \dfrac{ds}{dt} = 20 - 9.8t$

In this particular case, ds/dt is not only the gradient of the curve but also represents the rate of change of distance s with respect to time t. This in fact is velocity, so, if we write v for velocity, $v = ds/dt$ and hence

$$v = 20 - 9.8t$$

which corresponds to the standard equation for velocity with constant acceleration:

$$v = u + at$$

where u and a have the same values as before.

When differentiating an equation such as $v = u + at$, if it is necessary to make it quite clear that v and t are the variables we can specify that we intend to differentiate v with respect to t, which makes it quite clear that we are finding dv/dt. Also, when differentiating with respect to t (time) we are finding the rate at which a quantity is changing; for example, dv/dt represents the rate of change of velocity (which is acceleration).

C1.8 Alternative notation

We have seen earlier that if $y = x^3$ then $dy/dx = 3x^2$. We can replace y by the function itself and write $\dfrac{d(x^3)}{dx} = 3x^2$. It is common for differential coefficients to be expressed in this way. We use the symbol d/dx to denote the operation of differentiation with respect to x, just as we use $\sqrt{}$ to denote the operation of taking the square root.

Examples

a) $\dfrac{d}{dx}(2x^2 - 3x) = 4x - 3$

b) $\dfrac{d}{dt}(t^5 + 4t) = 5t^5 + 4$

If the relationship between a dependent variable and an independent variable is expressed in functional notation, we can use a different but equally common method of representing the differentiation operation. For example, we can write $y = 3x^2 + 4x + 2$ as $y = f(x)$ (see section B2.1). Differentiation is shown as $f'(x) = 6x + 4$.

Examples
a) $f(x) = 5x^{2.5}$, so $f'(x) = 12.5x^{1.5}$
b) $f(v) = 3v^2 - 2v$, so $f'(v) = 6v - 2$

C1.9 Differentiation of $\sin \theta$
In fig. C1.3, the top graph shows the familiar sine curve over a complete cycle covering values of x from 0 to 360° (0 to 2π radians). The gradient of the

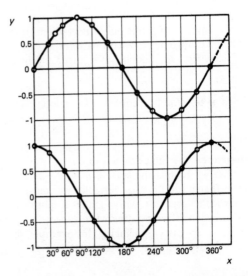

Fig. C1.3 Graph of $y = \sin x$

curve varies from +1 when $x = 0$ to zero when $x = 90°$, −1 when $x = 180°$, zero again at 270°, and back to +1 when $x = 360°$. The lower graph shows the way the gradient of the upper curve varies over this range. Noting that this lower curve is the graph of $y = \cos x$, we deduce that, when

100

$$y = \sin x$$

$$\frac{dy}{dx} = \cos x$$

Similarly, when

$$x = \sin \theta$$

$$\frac{dx}{d\theta} = \cos \theta$$

C1.10 Maximum or minimum

We introduced the concept of maxima and minima of curves in section B7.2. It is apparent from fig. C1.1 that the graph of $y = 4x - \frac{1}{2}x^2$ has a maximum when $x = 4$. This is the point on the curve at which the tangent is parallel to the x-axis, so the gradient of the curve is zero, i.e. $dy/dx = 0$.

Considering the graph of $y = \sin x$ in fig. C1.3, we see that the gradient is zero at the maximum (when $x = 90°$) and again at the minimum (when $x = 270°$). These are the only two points within this cycle of the sine wave at which $dy/dx = 0$.

Exercise C1

1 If P, Q, and R are the points $(1,1)$, $(2,8)$, and $(3,27)$ respectively on the graph of $y = x^3$, find the gradients of the tangents at the points P, Q, and R, and the gradient of the chord PR.

2 Differentiate $y = x^3$ from the first principles.

3 Differentiate $y = 1 + 2x^2$ from first principles.

4 Given that when $y = ax^n$, $dy/dx = ax^{n-1}$, find the value of dy/dx for each of the following:

(a) $y = 2x^3$ (b) $y = 5x^2 + 2$
(c) $y = (x + 3)^2$ (d) $y = 3x(x - 4)$

5 Given that $s = 36t + 4.9t^2$, find ds/dt.

6 If $s = 8.4 - 2t + 5t^2$, find the value of ds/dt when $t = 1.2$.

7 If $x = 2 + \sin \theta$, find the value of $dx/d\theta$ when $\theta = \pi/3$.

8 Find dy/dx when $y = 7.2 + 3x - x^2 + \sin x$.

9 Determine

(a) $\dfrac{d}{dx}(5x^3)$ (b) $\dfrac{d}{dt}(7t - 3t^2)$ (c) $\dfrac{d}{d\theta}(\theta + \sin \theta)$

10 (a) Find $f'(x)$ if $f(x) = 4x^{1.5} - 3/x^2 + 2$.
    (b) Find $f'(r)$ if $f(r) = Pr^{1.8}$, where P is constant.

11 Plot the graph of $y = 1 + 7x - x^2$ for values of x from 0 to 7. From the graph, find the gradient of the curve when $x = 2$. Find also the co-ordinates of the point on the curve at which the gradient is zero.

12 From the equation $s = 14.4t - 3.6t^2$, find the value of t for which ds/dt is zero.

C2 Integration

Starting with an equation expressing y in terms of x, we have seen how to differentiate y with respect to x to obtain the differential coefficient denoted by the symbol dy/dx. We shall now consider the reverse process, which is known as *integration*.

C2.1 Indefinite integration

An expression which becomes y when it is differentiated with respect to x is known as 'the *integral* of y with respect to x' and is denoted by the symbol

$$\int y \, dx$$

For example, if $y = x^3$, $dy/dx = 3x^2$

i.e. $\int 3x^2 \, dx = x^3$

or 'the integral of $3x^2$ with respect to x is x^3'.

However, if $y = x^3 + C$, where C is any constant, once again $dy/dx = 3x^2$. If, therefore, we are asked to integrate $3x^2$ with respect to x, i.e. to find $\int 3x^2 \, dx$, and we are not given any corresponding values of x and y, the result could be

$$y = x^3$$

or $\quad y = x^3 + 1$

or $\quad y = x^3 + \text{any constant}$

As we cannot give a definite relationship from which to calculate values of y for particular values of x, this result is known as an *indefinite integral* and is usually presented as

$$\int 3x^2 \, dx = x^3 + C$$

where C is known as the *arbitrary constant* and is added to cover the possibility of there having been a constant term in the relationship between x and y which was originally differentiated.

Differentiating $ax + \text{any constant}$ with respect to x, where a is also a constant, gives a, so

$$\int a \, dx = ax + C$$

whatever the value of the constant a.

C2.2 General form of the indefinite integral

Since $d(x^n)/dx = nx^{n-1}$, we can obtain x^n by differentiating

$$\frac{x^{n+1}}{n+1} + C$$

In reverse, therefore,

$$\int x^n \, dx = \frac{x^{n+1}}{n+1} + C$$

This result is true for all values of n, including negative and fraction values, *except* the case $n = -1$, when the denominator in the expression becomes zero. We will briefly mention this special case below, but the following examples illustrate the use of this result for some other values of n.

a) $\quad \int x^4 \, dx = \frac{x^5}{5} + C$

b) $\quad \int \frac{dx}{x^3} = \int x^{-3} \, dx = \frac{x^{-2}}{-2} + C = -\frac{1}{2x^2} + C$

c) $\quad \int \sqrt{x} \, dx = \int x^{\frac{1}{2}} \, dx = \frac{x^{\frac{3}{2}}}{\frac{3}{2}} + C = \frac{2}{3}\sqrt{x^3} + C$

It should be noted that, when a power of x is multiplied by a constant, that constant remains unchanged by the process of integration and can be placed outside the integration sign,

e.g. $\quad \int 6x^2 \, dx = 6 \int x^2 \, dx = \frac{6 \times x^3}{3} + C = 2x^3 + C$

Exception to the general form
The one exception to the general form above is the case where $n = -1$. In this case the integral is of logarithmic form and is unlikely to occur at this stage:

$$\int \frac{dx}{x} = \ln x + C$$

C2.3 Integration of a sum
When a sum of several terms is integrated, the result will be the sum of the integrals of the separate terms. Only one arbitrary constant is required.

Examples

a) $\quad \int (3x^2 - 8x + 4) \, dx = x^3 - 4x^2 + 4x + C$

b) $\quad \int (5x - \sqrt{x})^2 \, dx = \int (25x^2 - 10x^{\frac{3}{2}} + x) \, dx$

$$= \frac{25x^3}{3} - 4x^{\frac{5}{2}} + \frac{x^2}{2} + C$$

C2.4 Definite integrals – area under a curve

In fig. C2.1 we wish to find the area under the curve between the limits $x = a$ and $x = b$. Consider a strip of width δx (where δx is very small) at a distance x from the y-axis. The area of this narrow strip is approximately $y\delta x$ and the total area of the section required is given approximately by the sum of all such strips of area $y\delta x$, from $x = a$ to $x = b$.

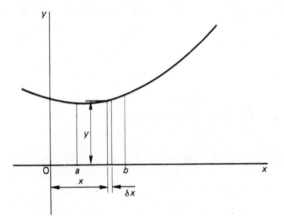

Fig. C2.1

Now the accuracy would be increased by increasing the number of strips, i.e. by reducing the value of δx. Taking this to the limit, the area is given accurately by an infinite number of strips of which the width δx has become the infinitesimal dx. For the sum of these strips we use the long s which has already been introduced as the symbol of integration, and the limits are placed at the head of this symbol. In this notation,

$$\text{area} = \int_a^b y \, dx$$

b and a are referred to as the upper and lower *limits* of the integral.

To calculate the area, the expression obtained after integration is evaluated first with the upper limit (b) substituted for x and then with the lower limit (a) substituted for x. The difference of these results (i.e. the evaluation using b minus the evaluation using a) is the area.

The process of evaluating the integral using the upper limit and then subtracting the evaluation using the lower limit is symbolised by enclosing the integral within square brackets with the upper and lower limits written outside.

i.e. if $y = x^2$

104

then $\displaystyle\int_a^b y\,dx = \int_a^b x^2\,dx$

$$= \left[\frac{x^3}{3} + C\right]_a^b = \left(\frac{b^3}{3} + C\right) - \left(\frac{a^3}{3} + C\right)$$

$$= \frac{b^3}{3} - \frac{a^3}{3} = \tfrac{1}{3}(b^3 - a^3)$$

Notice that for any particular curve the arbitrary constant will be the same for both evaluations and so does not enter into the result. An integral between limits is therefore known as a *definite* integral, and, as in the following example, it is not usual to introduce the arbitrary constant at all.

Example   Find the area between the curve $y = 2x^2 + 3x + 1$ and the x-axis between the limits $x = 1$ and $x = 2$.

$$\text{Area} = \int_1^2 (2x^2 + 3x + 1)\,dx$$

$$= \left[\frac{2x^3}{3} + \frac{3x^2}{2} + x\right]_1^2$$

$$= \left(\frac{16}{3} + 6 + 2\right) - \left(\frac{2}{3} + \frac{3}{2} + 1\right)$$

$$= 10.17 \text{ sq. units}$$

Areas above the axis come out positive, while parts of an area which are below the axis come out negative. This illustrates the necessity of checking whether or not a given curve crosses the axis within the given limits.

C2.5 Approximate methods for the evaluation of integrals

Various methods may be used for finding the areas of irregular figures, and these methods can be used to estimate the area under a particular curve. For Simpson's rule and for the trapezoidal rule (see sections D2.3 and D2.4) it is not even necessary to draw the corresponding curve. The interval between the limits of integration should be divided into an appropriate number of equally spaced smaller intervals and the corresponding ordinates calculated. The method is illustrated by example 1 in section D2.3. Integration gives this area as

$$\int_0^{12} (9x + 36 - x^2)\,dx = \left[\frac{9x^2}{2} + 36x - \frac{x^3}{3}\right]_0^{12}$$

$$= 648 + 432 - 576 = 504 \text{ sq. units}$$

C2.6 Integration of cos θ

We know that when $y = \sin\theta$, $dy/d\theta = \cos\theta$ (see section C1.8); consequently

$$\int \cos\theta \, d\theta = \sin\theta + C$$

Example Evaluate $\displaystyle\int_0^{\pi/3} 4\cos\theta \, d\theta$.

$$\int_0^{\pi/3} 4\cos\theta \, d\theta = [4\sin\theta]_0^{\pi/3}$$

$$= 4\sin(\pi/3) - 4\sin 0$$

$$= 3.464$$

Exercise C2

1 Integrate with respect to x (a) $4x$, (b) $8x^3$, (c) $x + 2$, (d) $6x^2 - 4x + 3$.

2 If $dy/dx = 9x^2 + 6x - 10$, find y, given that $y = 0$ when $x = 1$.

3 Given that $s = 0$ when $t = 0$, and $ds/dt = 49 - 9.8t$, find s in terms of t. Hence determine the value of s when $ds/dt = 0$.

4 Integrate the following functions with respect to x: (a) $9x^2 - 4x^3$, (b) $4(x - 2)$, (c) $1.8x^2 - 2.2x$.

5 Find (a) $\int 4\pi r^2 \, dr$, (b) $\int (4 - 10.8t^2) \, dt$.

6 Evaluate (a) $\displaystyle\int_1^2 (3x^2 - 4x) \, dx$, (b) $\displaystyle\int_0^2 (2 + 3x)^2 \, dx$

7 Evaluate $\displaystyle\int_0^{\pi/2} 2\cos\theta \, d\theta$.

8 Find the area enclosed between the x-axis and the curve $y = 7x - x^2 - 6$ by integrating between the limits of $x = 1$ and $x = 6$.

9 The gradient at any point on a certain curve is given by $3x^2 - 8x + 5$. Given that the curve passes through the point $(2,4)$ find its equation.

10 Show that the curve $y = 9 - x^2$ cuts the x-axis at $x = -3$ and $x = +3$. Plot the graph of the curve between these limits and use an approximate method to estimate the area enclosed under the curve. Check your answer by using integration to calculate the area.

D   Geometry, trigonometry, and vectors

D1 Volumes and surface areas

D1.1 Pyramids, cones, and spheres

We can calculate the volume or the surface area of a pyramid, cone, or sphere by using an appropriate formula from Table D1.1.

Title	Figure	Volume	Surface area
Square Pyramid		$\frac{1}{3}a^2h$	$2al + a^2$
Cone		$\frac{1}{3}\pi r^2 h$ $\frac{1}{12}\pi D^2 h$	$\pi rl + \pi r^2$ $\frac{1}{2}\pi Dl + \frac{1}{4}\pi D^2$
Sphere	Radius r Diameter D	$\frac{4}{3}\pi r^3$ $\frac{1}{6}\pi D^3$	$4\pi r^2$ πD^2

Table D1.1

Units of volume are cubic millimetres (mm^3), cubic centimetres (cm^3), and cubic metres (m^3), where

$$10^3\,mm^3 = 1\ cm^3 \quad \text{and} \quad 10^6\,cm^3 = 1\ m^3$$

When measuring the volume of fluid in a container, we often use the *litre* (l) which is 1000 cm^3, i.e. 1 ml = 1 cm^3.

Units of area are square millimetres (mm^2), square centimetres (cm^2), and square metres (m^2), where

$$10^2\,mm^2 = 1\ cm^2 \quad \text{and} \quad 10^4\,cm^2 = 1\ m^2$$

Example 1   Find the area of sheet metal needed to make a pyramid which has perpendicular height 80 mm and a square base of side 25 mm. Find also the volume of this pyramid.

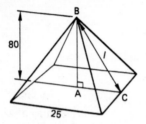

Fig. D1.1

The pyramid is shown in fig. D1.1. We use the formula

$$\text{area} = 2al + a^2$$

to calculate the total surface area, including that of the base.

To find the slant height l we can use triangle ABC, in which AB = 80 mm and AC = 12.5 mm. From the theorem of Pythagoras,

$$BC^2 = AB^2 + AC^2$$

$$\therefore \quad l = \sqrt{(80^2 + 12.5^2)}$$

$$= 80.97 \text{ mm}$$

The surface area is then

$$(2 \times 25 \times 80.97) + 25^2 = 4674 \text{ mm}^2$$

The volume of the pyramid is given by

$$V = \tfrac{1}{3}a^2h$$

$$= \tfrac{1}{3} \times 25^2 \times 80$$

$$= 1.667 \times 10^4 \text{ mm}^3, \text{ or } 16.7 \text{ cm}^3$$

Example 2   Find the mass of a solid steel cone with base radius 120 mm and height 250 mm, if the density of the steel is 8 g/cm³.

The volume of the cone is given by

$$V = \tfrac{1}{3}\pi r^2h$$

$$= \tfrac{1}{3} \times \pi \times 120^2 \times 250$$

$$= 3.770 \times 10^6 \text{ mm}^3$$

$$= 3770 \text{ cm}^3$$

The mass of the cone is therefore

$$3770 \times 8 = 30\,160\,g, \text{ or } 30.16\,kg$$

Example 3   How many litres of water can be stored in a spherical tank
5.5 m in diameter?

The volume of a sphere with diameter D is calculated using

$$V = \tfrac{1}{6}\pi D^3$$

Substituting $D = 5.5\,m$ gives

$$V = \tfrac{1}{6}\pi \times 5.5^3 = 87.11\,m^3$$

Now, $1\,m^3 = 10^6\,cm^3$ and 1 litre $= 1000\,cm^3$, therefore $1\,m^3 = 1000$ litres.
The volume (or *capacity*) of the tank is therefore 87 110 litres.

We see from Table D1.1 that the volume of a square pyramid is $\tfrac{1}{3} \times$
base area × perpendicular height. This is true even when the base is not
square. For example the volume of the pyramid in fig. D1.2 is found as
follows:

$$\text{volume} = \tfrac{1}{3} \times (5 \times 10) \times 7.5 = 125\,m^3$$

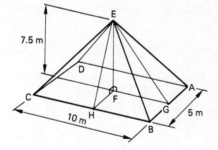

Fig. D1.2

To find the total surface area we must note that the triangular sides of this
pyramid do not have the same area, and so we cannot use the formula for a
square-based pyramid. To find the area of each triangular side we must first
calculate the slant height.

In triangle EHF, using the theorem of Pythagoras,

$$EH^2 = EF^2 + FH^2$$

$$\therefore \quad EH = \sqrt{(7.5^2 + 2.5^2)} = \sqrt{62.5}$$

$$= 7.91\,m$$

The area of triangle BEC is then

$\frac{1}{2} \times 10 \times 7.91 = 39.53 \, \text{m}^2$

Similarly, in triangle EGF,

$EG^2 = 5^2 + 7.5^2$

$\therefore \quad EG = 9.01 \, \text{m}$

The area of triangle AEB $= \frac{1}{2} \times 5 \times 9.01$

$= 22.53 \, \text{m}^2$

Base area ABCD $= 10 \times 5 = 50 \, \text{m}^2$

$\therefore \quad$ total surface area $= (2 \times 39.53) + (2 \times 22.53) + 50$

$= 174.1 \, \text{m}^2$

D1.2 The frustum

In section D1.1 we saw that a pyramid is a figure with a square or rectangular base and triangular sides which form a pointed top. If a section is removed from the top of a pyramid, the remaining shape is called a *frustum* (plural *frusta*).

Figure D1.3(a) shows a frustum of a pyramid, and we see that the top is the same shape as the base but with reduced dimensions, provided the cut is made parallel to the base. The sides are *trapezoidal* in shape.

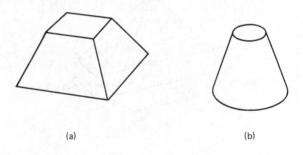

(a)   (b)

Fig. D1.3 Frusta of a pyramid and of a cone

In fig. D1.3(b) we see a frustum of a cone with a circular top. The section removed is of course a small cone, and the volume and surface area of the remaining frustum can be found by subtraction. Formulae exist to simplify these calculations and they are given in the next section.

D1.3 Volumes and surface areas of frusta

Table D1.2 gives formulae for the volume and surface area of the frustum of a pyramid and of a cone, with formulae for a segment of a sphere.

Title	Figure	Volume	Total surface area
Frustum of a square pyramid	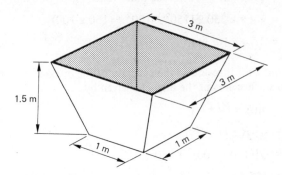	$\frac{1}{3}h(A^2 + a^2 + Aa)$	$2l(A + a) + A^2 + a^2$ where: $l = \sqrt{\left(\frac{A - a}{2}\right)^2 + h^2}$
Frustum of a cone		$\frac{1}{3}\pi h(R^2 + r^2 + Rr)$	$\pi(R + r)l + \pi(R^2 + r^2)$ where $l = \sqrt{(R - r)^2 + h^2}$
Segment of a sphere		$\frac{\pi h}{6}(3r^2 + h^2)$ or $\frac{\pi h^2}{3}(3R - h)$	$4\pi Rh - \pi h^2$ $= 2\pi Rh + \pi r^2$

Table D1.2 Volumes and surface area of frusta

Example 1   Find the volume of the storage tank shown in fig. D1.4 and the area of steel plate used to make it with a base but no top.

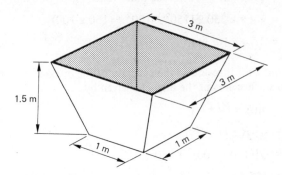

Fig. D1.4

The volume of a frustum of a pyramid is given by

$$V = \tfrac{1}{3}h(A^2 + a^2 + Aa)$$

where $h = 1.5\,\text{m}$ $A = 3\,\text{m}$ and $a = 1\,\text{m}$

111

Substituting these values gives

$$V = \frac{1.5}{3}(9 + 1 + 3)$$

$$= 6.5\,\text{m}^3$$

The total surface area without a top is given by

$$\text{area} = 2l(A + a) + a^2$$

where $\quad l = \sqrt{\left[\left(\frac{A-a}{2}\right)^2 + h^2\right]} = \sqrt{\left[\left(\frac{3-1}{2}\right)^2 + 1.5^2\right]}$

$$= 1.803\,\text{m}$$

$\therefore \quad$ area of plate $= 2 \times 1.803 \times (3 + 1) + 1$

$$= 15.42\,\text{m}^2$$

Example 2   Calculate the capacity in litres of a plastic bucket with base diameter 200 mm, top diameter 300 mm, and height 350 mm. Find also the mass of the bucket, assuming uniform thickness, if each square centimetre of plastic has mass 0.32 g.

We can use the formula for the volume of a frustum of a cone:

$$\text{volume} = \tfrac{1}{3}\pi h(R^2 + r^2 + Rr)$$

where $\quad R = 150\,\text{mm} \qquad r = 100\,\text{mm} \quad$ and $\quad h = 350\,\text{mm}$

so that $\quad V = \tfrac{1}{3} \times \pi \times 350 \times [150^2 + 100^2 + (150 \times 100)]$

$$= 1.74 \times 10^7\,\text{mm}^3$$

Now, $10^6\,\text{mm}^3 = 1$ litre and so the capacity of the bucket $= 17.4$ litres.
    The total surface area of the bucket is given by

$$\text{area} = \pi[(R + r)l + r^2]$$

where $\quad l = \sqrt{[(R - r)^2 + h^2]}$

$$= \sqrt{[(150 - 100)^2 + 350^2]}$$

$$= 353.6\,\text{mm}$$

$\therefore \quad$ area $= \pi[(150 + 100) \times 353.6 + 100^2]$

$$= 3.091 \times 10^5\,\text{mm}^2$$

$$= 3091\,\text{cm}^2$$

$\therefore \quad$ mass of the bucket $= 3091 \times 0.32\,\text{g} = 989\,\text{g}$

Example 3   Find the volume of glass in a lens which is a segment of a 120 mm diameter sphere and has a maximum thickness of 7.5 mm.

The formula for the volume of a segment of a sphere appropriate here is

$$\text{volume} = \frac{\pi h^2}{3}(3R - h)$$

where $h = 7.5\,\text{mm}$ and $R = 120/2 = 60\,\text{mm}$

$\therefore$ $\text{volume} = \dfrac{\pi \times 7.5^2}{3}(180 - 7.5)$

             $= 1.016 \times 10^4\,\text{mm}^3$

D1.4 Component parts of composite shapes

Calculating composite areas and volumes of engineering components can often be made simpler by identifying rectangles, triangles, circles, cylinders, and spheres, together with other basic shapes which make up the component. The total area or volume is then found by summation of these individual shapes. Figure D1.5 shows some composite areas and the basic shapes which could conveniently be used.

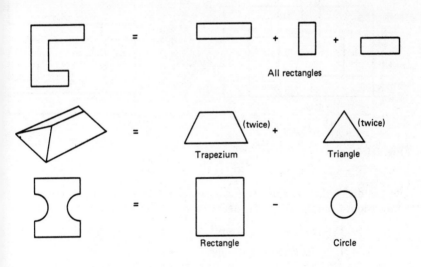

Fig. D1.5 Basic shapes from some composite areas

In fig. D1.6 there are some objects which could be divided into common shapes so that their volumes could be calculated. While it may not be possible to obtain highly accurate answers this way, it is often useful to be able to give an approximate volume, to determine raw-material requirements for example.

113

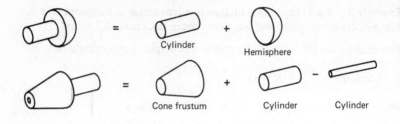

Fig. D1.6

D1.5 Total areas and volumes of composite figures

Having identified suitable basic elements in a composite shape as described in section D1.4, the total area or volume can be determined by summation.

Example 1   Find the cross-sectional area of the extruded component in fig. D1.7.

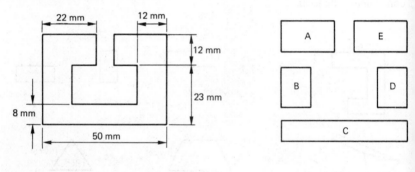

Fig. D1.7   Fig. D1.8

We can divide the area into the five rectangles shown in fig. D1.8. Using dimensions from fig. D1.7, the areas of the symmetrical cross-sections are

$$A = E = 22 \times 12 = 264 \, mm^2$$
$$B = D = 12 \times 15 = 180 \, mm^2$$
$$C \quad\;\; = 50 \times 8 \;\; = 400 \, mm^2$$

The total cross-sectional area is therefore

$$(2 \times 264) + (2 \times 180) + 400 = 1288 \, mm^2$$

Note that there are other ways we could have divided the shape into rectangles. One alternative would be to calculate the area of the complete outer rectangle and subtract the missing areas, giving

114

$$\text{area} = (50 \times 35) - (6 \times 12) - (26 \times 15)$$
$$= 1288 \text{ mm}^2$$

Example 2 Find the area of the simplified boat deck in fig. D1.9.

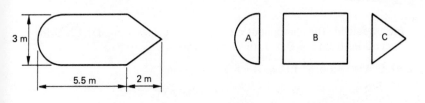

Fig. D1.9   **Fig. D1.10**

Dividing the boat deck into basic shapes A, B, and C as shown in fig. D1.10, we can calculate the total area.

Using dimensions from fig. D1.9, the individual areas are

$$A = \tfrac{1}{2} \times \frac{\pi \times 9}{4} = 3.534 \text{ m}^2$$

$$B = 4 \times 3 = 12 \text{ m}^2$$

$$C = \tfrac{1}{2} \times 3 \times 2 = 3 \text{ m}^2$$

The area of the boat deck is therefore 18.53 m².

Example 3 Determine the volume of brass in the electrical connector component shown in fig. D1.11.

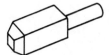

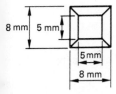

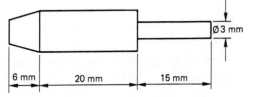

Fig. D1.11

115

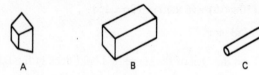

A   B   C

Fig. D1.12

First we divide the component into the three shapes in fig. D1.12. Using dimensions from fig. D1.11, the volumes are

$$A = \tfrac{1}{3} \times 6(8^2 + (8 \times 5) + 5^2) = 258 \, mm^3$$

$$B = 8 \times 8 \times 20 \qquad\qquad = 1280 \, mm^3$$

$$C = \frac{\pi \times 3^2}{4} \times 15 \qquad\qquad = 106.0 \, mm^3$$

The total volume of the component is therefore $1644 \, mm^3$.

D1.6 Volume of a prism

To find the volume of a cylinder we simply multiply the area of the circular end by its length. This method can be applied to all *prisms*, i.e. objects with a constant cross-section:

volume of a prism = area of cross-section × length

Example 1 Find the volume of a 2 m length of hexagonal bar which is 20 mm across flats.

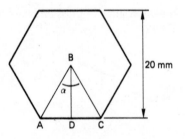

Fig. D1.13

Figure D1.13 shows the cross-section of the bar. From the symmetry of the figure, angle $\alpha = 360°/6 = 60°$, and, in triangle DBC,

$$DC = 10 \tan 30°$$

$$= 5.774 \, mm$$

116

The area of triangle ABC is therefore

$\frac{1}{2} \times$ AC $\times$ BD $= 5.774 \times 10$

$\qquad = 57.74 \, \text{mm}^2$

The total cross-sectional area $= 6 \times 57.74 = 346.4 \, \text{mm}^2$

From this, the volume of a 2 m length of hexagonal bar is given by

volume $=$ area of cross-section $\times$ length

$\qquad = 346.4 \times 2000$

$\qquad = 6.93 \times 10^5 \, \text{mm}^3$

Example 2 Concrete drainage channels with the cross-section shown in fig. D1.14 are cast in 1500 mm lengths. Calculate the mass of each of these castings if the density of concrete is 2.2 g/cm³.

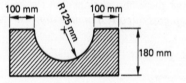

Fig. D1.14

We first calculate the cross-sectional area of the channel, as follows.

Area $= (450 \times 180) - (\frac{1}{2} \times \pi \times 125^2)$

$\qquad = 5.646 \times 10^4 \, \text{mm}^2$

The volume of one casting is then

cross-sectional area $\times$ length $= 5.646 \times 10^4 \times 1500$

$\qquad\qquad\qquad\qquad = 8.469 \times 10^7 \, \text{mm}^3$

$\qquad\qquad\qquad\qquad = 8.469 \times 10^4 \, \text{cm}^3$

Mass $=$ volume $\times$ density

$\qquad = 8.469 \times 10^4 \times 2.2$

$\qquad = 1.863 \times 10^5 \, \text{g}$

i.e. the mass of each casting is 186.3 kg.

D1.7 Areas and volumes of similar shapes

Areas
The sides of square B in fig. D1.15 are twice as long as those of square A. We can see that four of the smaller squares could be put together to make a square

117

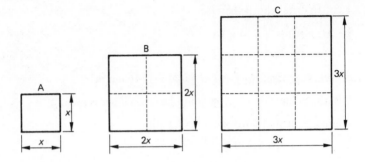

Fig. D1.15

exactly the same shape as B. The area of square B is thus four times that of square A. Similarly, square C, which has sides three times as long as those of square A, has nine times the area of square A.

We can deduce from these observations that *the areas of similar shapes are proportional to the squares of corresponding dimensions.* Using the dimensions in fig. D1.15,

$$\text{area A} = x \times x = x^2$$

$$\text{area B} = 2x \times 2x = 4x^2$$

$$\text{area C} = 3x \times 3x = 9x^2$$

Note also that the area of square C is 2.25 times that of B, since its dimensions are one and a half times those of B and $(1.5)^2 = 2.25$.

Example   The plans of a house drawn to a scale of 1 to 50 show a rectangular room with plan measurements 72 mm by 46 mm. What is the true area of the room in square metres?

The area shown on the plan $= 72 \times 46 = 3312 \text{ mm}^2$

$\therefore$ the true area of the room $= 3312 \times (50)^2 = 8.28 \times 10^6 \text{ mm}^2$

i.e. the area of the room is 8.28 m^2.

Volumes

By comparing the volumes of the three square-sided blocks in fig. D1.16, we can deduce that *the volumes of similar bodies are proportional to the cubes of corresponding dimensions.*

It would take eight of the smallest blocks with sides of length x to construct the block with sides of length $2x$. To make a block with sides of length $3x$ we would require 27 of the smallest blocks.

$$\text{Volume of block A} = x \times x \times x = x^3$$

$$\text{Volume of block B} = 2x \times 2x \times 2x = 8x^3$$

118

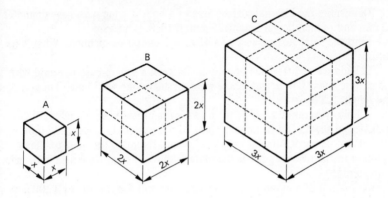

Fig. D1.16

Volume of block C $= 3x \times 3x \times 3x = 27x^3$

To make a block with sides of length $10x$, we would need $10^3 = 1000$ of the smallest blocks. Note that the volume of block C is 3.375 times as large as the volume of block B. This can easily be seen if we count the number of small blocks with side x in each of the blocks B and C. To avoid counting we can obtain the result by simply cubing the ratio of the dimensions of the two blocks:

$$\frac{\text{volume of block C}}{\text{volume of block B}} = \left(\frac{3x}{2x}\right)^3 = (1.5)^3 = 3.375$$

Example A model of a vehicle is made to a scale of 1 to 75 for wind-tunnel tests. If the volume of the vehicle is $6.33 \times 10^9 \text{ mm}^3$, what will be the volume of the model?

The volumes of the vehicle and the model will be in proportion to the cubes of their corresponding dimensions:

$$\frac{\text{dimensions of the vehicle}}{\text{dimensions of the model}} = \frac{75}{1}$$

$\therefore \qquad \dfrac{\text{volume of the vehicle}}{\text{volume of the model}} = (75)^3 = 4.22 \times 10^5$

$$\text{volume of the model} = \frac{6.33 \times 10^9}{4.22 \times 10^5} = 1.50 \times 10^4 \text{ mm}^3$$

i.e. the volume of the model is $1.50 \times 10^4 \text{ mm}^3$.

Exercise D1
1 Calculate the total surface area of a pyramid with base 36 mm square and height 50 mm.

2 Determine the internal surface area of a hollow cone with base diameter 12 mm and perpendicular height 25 mm, if there is no base.

3 Find the total surface area of a 48 mm diameter hemisphere. What is its volume?

4 Calculate the difference in volume between a square-base pyramid with 1.5 m base and 2.8 m height, and a right circular cone with base diameter 3 m and height 2.8 m.

5 A hollow pyramid is made from two pairs of isosceles triangles cut from epoxy resin sheet. The rectangular base of the pyramid measures 42 mm by 30 mm and its perpendicular height is 75 mm. Estimate the mass of the pyramid to the nearest gram if the material has constant thickness with mass 0.45 g/cm².

6 A frustum of a pyramid has a square base and top measuring 1200 mm and 700 mm respectively. If the perpendicular height is 550 mm, calculate the volume and total surface area, including base and top.

7 The outer surface of a satellite forms a frustum of a cone with large and small diameters 950 mm and 600 mm, and with slant height 750 mm. This curved area is to be covered with solar cells each with an area of 400 mm². Determine the approximate number of cells required.

8 Two identical spherical caps are cut from opposite sides of a 36 mm diameter steel ball to leave parallel plane surfaces 22 mm apart. Calculate the remaining volume.

9 Calculate the shaded areas in fig. D1.17.

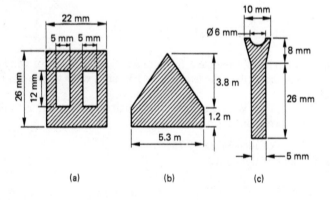

Fig. D1.17

10 Find the volumes of the components shown in fig. D1.18.

11 Figure D1.19 shows the internal dimensions of a loudspeaker enclosure. Calculate the depth x to the nearest millimetre, if the volume of air surrounding the speaker is to be 0.1 m³.

12 The cross-section of a bar is an equilateral triangle of side 18 mm. Find the volume of 350 mm of this bar.

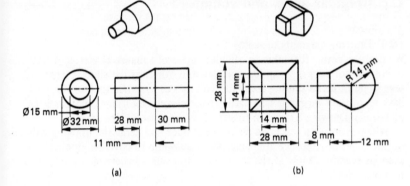

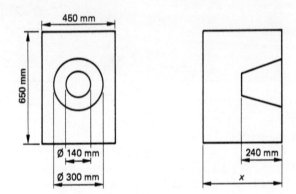

Fig. D1.18

Fig. D1.19

13 Find the mass of 1 km of 2 mm diameter copper wire if the density of copper is 8900 kg/m³.

14 Air flows through a 190 mm by 250 mm rectangular air-conditioning duct at a velocity of 3 m/s. Find the volume flow in cubic metres per hour.

15 The area of a forest on a map drawn to a scale 1 to 50 000 is 750 mm². What is the true area of the forest in km²?

16 The mass of a solid steel engineering component is 135 g. If the dimensions of the component were reduced uniformly by 20%, what would be its new mass?

D2 Irregular areas and volumes

D2.1 Drawing diagrams to scale

We can use simple formulae to calculate areas of squares, circles, and triangles; or volumes of cubes, pyramids, spheres, and other *regular* shapes. There are several methods of finding areas and volumes of *irregular* shapes which do not have an outline we can define easily (for example, by using a mathematical equation). Drawing a diagram to scale is often a useful first step.

When choosing a scale, we need to consider how large the drawing will be on the sheet of paper we are using. We should aim to cover as much of the sheet as possible, while at the same time choosing a scale which is convenient to work with.

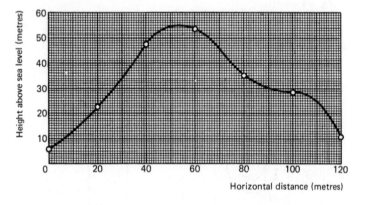

Fig. D2.1 Cross-section of a hill

Figure D2.1 shows a drawing of the cross-section of a hill, made using the following table of heights above sea level at intervals of 10 m measured horizontally from a reference point at the foot of the hill.

Horizontal distance (m)	0	20	40	60	80	100	120
Height above sea level (m)	5.5	22.1	46.4	52.2	34.5	27.6	10.1

The scale used here is one small square to represent one metre, both horizontally and vertically. We can only estimate the true height of the hill between the measured points by joining these points with a smooth curve.

To evaluate the area of the cross-section, we can count squares on our drawing. Each small square represents one square metre, so a large square is equivalent to $100\,\text{m}^2$. Counting first the number of complete large squares and then the remaining small squares gives the following area:

29   complete large squares

825   small squares

$$\text{Total area} = 2900 + 825$$
$$= 3725 \, \text{m}^2$$

Errors in this value arise from deviations of the outline of our scale drawing from the true outline of the hill, and from the way we use our judgement to include or exclude incomplete small squares in the total. A better estimate can be obtained by making more height measurements and drawing a diagram to a still larger scale; however this would make counting squares even more laborious.

D2.2 Using a planimeter

A *planimeter* is an instrument for measuring plane areas directly from a scale drawing. There are various types, but basically each has a pointer which is moved around the boundary of the area to provide automatically a measure of the area, which is usually recorded on a graduated wheel or cylinder. Figure D2.2 shows a planimeter being used to measure the area of the hill cross-section in fig. D2.1.

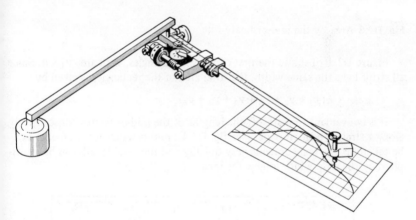

Fig. D2.2 Using a planimeter to measure area

Before use, the planimeter should be calibrated carefully according to the instructions supplied with the instrument. This can be done by measuring a known area such as a square of side 10 cm. With the planimeter scale set at 1:1, the area traced should be recorded as 100 cm² exactly. Adjustments can be made to the vernier position on the tracer bar until this result is obtained.

D2.3 The mid-ordinate and trapezoidal rules

In sections D2.1 and D2.2 we saw how to find an irregular area by counting squares and using a planimeter. The mid-ordinate and trapezoidal rules provide methods of obtaining an approximate value for such areas by calculation.

The mid-ordinate rule

To find the area below line AB in fig. D2.3(a), we construct vertical lines which divide the area into a number of equal strips. In this case we have used six strips, but the answer will be more accurate if a larger number are used. The *mid-ordinate* of each strip is drawn ($y_1, y_2, \ldots, y_6$) and the areas of six rectangles with the heights of these mid-ordinates are added to give the total area.

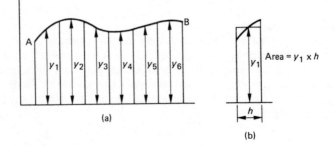

Fig. D2.3 Area by the mid-ordinate rule

Figure D2.3(b) shows the first of the six rectangles, with area $y_1 \times h$. Since all strips have the same width, the total area of the rectangles is given by

$$\text{area} = h(y_1 + y_2 + y_3 + y_4 + y_5 + y_6)$$

It is convenient to mark off the lengths of the mid-ordinates consecutively along a straight edge of paper (see fig. D2.4); in this way, measurements can be taken off the scale drawing very quickly and the total length can be multiplied by the strip width to give the area.

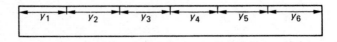

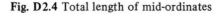

Fig. D2.4 Total length of mid-ordinates

The trapezoidal rule

The area of any trapezium is given by half the sum of the parallel sides times the perpendicular distance between them. The area is first divided into strips by equally spaced ordinates (see fig. D2.5(a)), as for the mid-ordinate rule. The area of each strip is assumed to be equal to that of the trapezium formed when the ends of these ordinates are joined as shown in fig. D2.5(b).

The total area PQRS is given by

$$\text{area} = \frac{h}{2}(y_1 + y_2) + \frac{h}{2}(y_2 + y_3) + \frac{h}{2}(y_3 + y_4) + \ldots + \frac{h}{2}(y_6 + y_7)$$

124

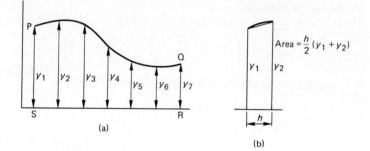

Fig. D2.5 Area by the trapezoidal rule

which simplifies to give

$$\text{area} = h[\tfrac{1}{2}(y_1 + y_7) + y_2 + y_3 + y_4 + y_5 + y_6]$$

where h is the width of each strip.

It is convenient to remember this in the form

$$\text{area} = \text{strip width} \times (\text{average of first and last} + \text{sum of all other}$$
$$\text{ordinates})$$

Greater accuracy is obtained by increasing the number of strips, as with the mid-ordinate rule, but if the curvature of the boundary is continually one way the estimate of the area can only be approximate.

Example 1   Find the area enclosed by the curve $y = 9x + 36 - x^2$ and the positive x- and y-axes, using the mid-ordinate rule and the trapezoidal rule with six strips in each case.

Figure D2.6 shows the area divided into six strips.

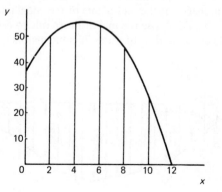

Fig. D2.6

125

By the mid-ordinate rule
The mid-ordinates are values of y at $x = 1, 3, 5, \ldots, 11$ as follows:

x	1	3	5	7	9	11
y	44	54	56	50	36	14

Area = strip width x sum of mid-ordinates

= $2(44 + 54 + 56 + 50 + 36 + 14)$

= 512 sq. units

By the trapezoidal rule
The ordinates required are values of y at $x = 0, 2, 4, \ldots, 12$:

x	0	2	4	6	8	10	12
y	36	50	56	54	44	26	0

Area = strip width x (average of first and last + sum of all other ordinates)

$$= 2 \times \left[\frac{(36 + 0)}{2} + 50 + 56 + 54 + 44 + 26 \right]$$

= 496 sq. units

Note that we would expect the answer obtained using the trapezoidal rule to be smaller than the correct value, since the curve bends one way continuall and all trapezoids will lie below the curve. The correct answer is in fact 504 square units (see section C2.5), and so the answer we obtained using the mid-ordinate rule is too large in this case. With a tortuous boundary the small errors would tend to cancel out.

Example 2 A river is 10 m wide and the average flow of water is 3 m/s. Depths of water in metres at the mid-points of ten equal intervals across the river are shown in fig. D2.7. Use the mid-ordinate rule to estimate the area of cross-section and hence the flow of water in cubic metres per second.

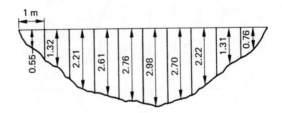

Fig. D2.7

By the mid-ordinate rule,

  area = strip width x sum of mid-ordinates

   = 1 x (0.55 + 1.32 + 2.21 + 2.61 + 2.76 + 2.98 + 2.70 + 2.22 +

   1.31 + 0.76)

   = 19.42 m²

Flow of water = area x velocity

   = 19.42 x 3

   = 58.26 m³/s

i.e. the flow of water in the river is approximately 58 m³/s.

Example 3   A vehicle decelerates from a velocity of 25 m/s and comes to rest in 30 s. The following table gives velocities at 5 s intervals.

Time (s)	0	5	10	15	20	25	30
Velocity (m/s)	25	22.5	20	17	14.5	9.5	0

Use the trapezoidal rule to find the area below the velocity–time curve and hence the distance travelled by the vehicle before coming to rest.

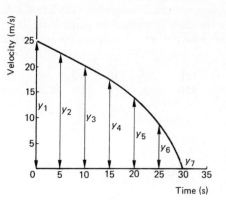

Fig. D2.8

Figure D2.8 shows the velocity–time curve with given ordinates at 5 s intervals. By the trapezoidal rule,

  area = $h \times [\frac{1}{2}(y_1 + y_7) + y_2 + y_3 + y_4 + y_5 + y_6]$

   = $5 \times [\frac{1}{2}(25 + 0) + 22.5 + 20 + 17 + 14.5 + 9.5]$

   = 480 m

127

Note that the units of this area are metres, since

velocity (m/s) x time (s) ≡ distance (m)

i.e. the distance travelled by the vehicle before coming to rest is approximately 480 m.

D2.4 Simpson's rule

Simpson's rule provides the most accurate of the formulae for area approximation, and for many regular curves it is exact. It is obtained by combining the trapezoidal formula with the mid-ordinate formula, and is expressed as

$$\text{area} = \frac{h}{3}(A + 2B + 4C)$$

where h = strip width

A = sum of first and last ordinate

B = sum of remaining odd ordinates

C = sum of all even ordinates

Note: this form of Simpson's rule requires an *odd number of ordinates*, i.e. an even number of strips.

Example   The force F newtons required to move a load varies with the distance s metres through which the load is moved in the direction of the force as shown in the table below and the graph in fig. D2.9. Using Simpson's rule, find the area under the curve and hence the total work done in moving the load.

s (m)	0	10	20	30	40	50	60
F (N)	840	810	730	620	510	420	360
	y_1	y_2	y_3	y_4	y_5	y_6	y_7

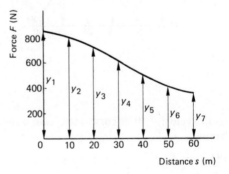

Fig. D2.9

128

The table of values has an odd number of ordinates, which is a necessary condition. By Simpson's rule,

$$\text{area} = \frac{h}{3}(A + 2B + 4C)$$

where $h = 10\,\text{m}$

$A = 840 + 360 = 1200\,\text{N}$

$B = 730 + 510 = 1240\,\text{N}$

$C = 810 + 620 + 420 = 1850\,\text{N}$

$\therefore \quad \text{area} = \frac{10}{3}(1200 + 2480 + 7400)$

$= 3.693 \times 10^4\,\text{N m}$

and, since 1 joule = 1 newton x 1 metre, the total work done by the force is 37 kJ.

D2.5 Volumes by Simpson's rule
Replacing ordinates by areas of cross-section enables volumes to be found by Simpson's rule.

Example Use Simpson's rule to find the approximate capacity in litres of an aircraft fuel tank, using the diameters of its circular cross-section at 500 mm intervals given in fig. D2.10.

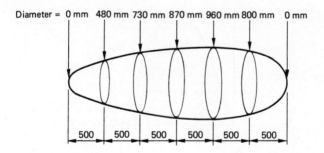

Diameter = 0 mm   480 mm   730 mm   870 mm   960 mm   800 mm   0 mm

500   500   500   500   500   500

Fig. D2.10

Areas of cross-section can be calculated from the given diameters as follows:

Diameter (mm)	0	480	730	870	960	800	0
Area (m²)	0	0.181	0.419	0.594	0.724	0.503	0

By Simpson's rule,

$$\text{volume} = \frac{h}{3}(A + 2B + 4C)$$

where h = 0.500 m (i.e. distance between cross-sections)

A = 0 + 0 = 0 m (i.e. sum of first and last areas)

B = 0.419 + 0.724 = 1.143 m (i.e. sum of remaining odd areas)

C = 0.181 + 0.594 + 0.503 = 1.278 m (i.e. sum of all even areas)

$\therefore$ volume = $\dfrac{0.500}{3}$ (0 + 2.286 + 5.112)

= 1.233 m³

and   1 m³ = 1000 litres

i.e. the fuel-tank capacity is approximately 1230 litres.

D2.6 Accuracy of numerical results

All the methods for finding areas and volumes described in sections D2.1 to D2.5 usually give approximate answers. Their accuracy depends upon how well we can make measurements and draw diagrams to scale when counting squares or using a planimeter, and upon how many strips or sections we use with Simpson's rule and the mid-ordinate and trapezoidal rules. The following table gives an example of how the value of an area given by these last three rules can become more accurate when the number of strips is increased. The area below the curve $y = \sin x$ between $x = 0°$ and $x = 180°$ which is illustrated in fig. D2.11 is exactly 2 square units. The table gives the area we would calculate using the three rules with 2, 4, and 6 strips.

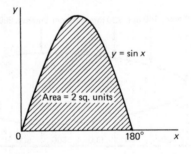

Fig. D2.11 Area below one loop of a sine curve

Rule used	Number of strips					
	2		4		6	
	Area	% error	Area	% error	Area	% error
Mid-ordinate	2.221	+11.05	2.052	+2.60	2.023	+1.15
Trapezoidal	1.571	−21.45	1.896	−5.20	1.954	−2.30
Simpson's	2.094	+4.70	2.005	+0.25	2.001	+0.05

130

There is no sure way of deciding exactly how accurate our approximation of an irregular area or volume will be: it is necessary to look at the quality of the data and the limitations of the formula in each case, and to state numerical results to an accuracy consistent with these sources of error. Always compare your answer with a rough estimate of the area or volume; many simple arithmetic mistakes can be avoided in this way.

These numerical methods for the determination of areas are suitable for use with programmable calculators and computers. Large numbers of strips can then be used to give very accurate answers.

D2.7 The mean value of common waveforms

The mean (or average) value of a curve above an axis can be found if we know the area enclosed below the curve between two points on the axis:

$$\text{mean value} = \frac{\text{area below the curve}}{\text{length of base}}$$

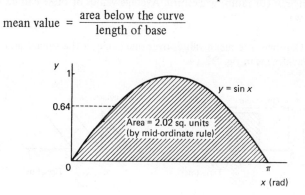

Fig. D2.12 $y = \sin x$ (half-wave)

In fig. D2.12, the curve $y = \sin x$ is shown between $x = 0$ and π radians (or $180°$). This is a *half wave*, or half of one complete cycle of the waveform.

$$\text{Mean value} = \frac{\text{area below curve between 0 and } \pi \text{ radians}}{\pi}$$

In section D2.6, using the mid-ordinate rule with six strips, we found this area to be approximately 2.02 square units;

$$\therefore \quad \text{mean value} = \frac{2.02}{3.14} = 0.64 \text{ sq. units}$$

The exact mean value is in fact 0.6366 sq. units and so the error in our answer is about $\frac{1}{2}\%$.

In fig. D2.13 we see a *full wave* of $y = \sin x$ drawn between $x = 0$ and 2π radians. The total area bounded by the curve and the x-axis is zero (since the area below the axis is negative and equal in size to that above the axis),

$$\therefore \quad \text{mean value for a full wave} = \frac{0}{2\pi} = 0$$

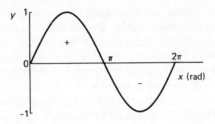

Fig. D2.13 $y = \sin x$ (full-wave)

Other waveforms, commonly used in electrical work, are the square-wave and the triangular (or ramp) waveform. Average values of these can be found in the same way.

Example Compare the mean values over one cycle of the square and triangular waveforms in fig. D2.14.

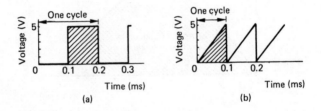

Fig. D2.14

Square-wave

$$\text{Mean value} = \frac{\text{area under curve}}{\text{base length}}$$

Time for one cycle = 0.2 ms and the shaded area in fig. D2.14(a) = $5\,V \times 0.1\,ms = 0.5\,V\,ms$

$\therefore$ mean value $= \dfrac{0.5}{0.2} = 2.5\,V$

Ramp-wave
One cycle of the ramp-wave is completed in 0.1 ms and the area below the curve in fig. D2.14(b) (shown shaded) = $\frac{1}{2} \times 0.1\,ms \times 5\,V = 0.25\,V\,ms$

$\therefore$ mean value $= \dfrac{0.25}{0.1} = 2.5\,V$

i.e. the mean value is the same for both waveforms.

Exercise D2

1 The following table gives measurements at 100 m intervals measured horizontally taken during a geological survey of depths below sea level of the upper and lower surfaces of an oil deposit. Draw to scale a cross-section of this oil field and, by counting squares or using a planimeter, estimate the area involved.

Horizontal measurement (m)	0	100	200	300	400	500	600	700
Depth below sea level of upper surface of oil (m)	170	130	130	140	130	125	135	165
Depth of lower surface (m)	175	220	245	270	280	270	250	170

2 State clearly the mid-ordinate rule, the trapezoidal rule, and Simpson's rule for finding irregular areas, and say which you think is the most accurate.

3 The depth of a river 32 m wide is measured at the eight points given below. Use the mid-ordinate rule to find the area of cross-section of the river.

Perpendicular distance from bank (m)	2	6	10	14	18	22	24	30
Depth of river (m)	2.1	3.3	5.8	6.7	7.7	7.2	4.9	1.2

4 Plot a graph of y against x using the table of values below, and find the area below this curve between $x = 0$ mm and $x = 1.4$ mm, using a planimeter or by counting squares. Compare this value with the area calculated using the mid-ordinate rule, with seven mid-ordinates measured from the graph.

x (mm)	0	0.2	0.4	0.6	0.8	1.0	1.2	1.4
y (mm)	1.00	1.20	1.38	1.54	1.66	1.76	1.83	1.89

5 The velocity of a vehicle is measured every 2 s as it accelerates from rest. From the following values, plot a graph of velocity against time and use it to estimate the velocity after 5 s. Find the distance travelled by the vehicle in 10.0 s by using the trapezoidal rule to calculate the area below your curve.

Time (s)	0.0	2.0	4.0	6.0	8.0	10.0
Velocity (m/s)	0.0	2.01	4.11	6.37	8.88	11.75

6 Figure D2.15 shows the outline of a forest on a map. Use the trapezoidal rule to calculate the area, and find the true area of the forest, in square kilometres, if the scale of the map is 1 : 50 000.

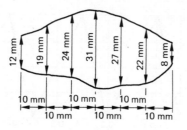

Fig. D2.15

133

7 Use Simpson's rule to find the area under a graph of force against distance from the following results, which were taken during an experiment in which a piston was moved 50 mm in a cylinder. Express your answer as work done in joules.

Distance moved by piston (mm)	0.0	7.5	15.0	22.5	30.0	37.5	45.0
Force (N)	10.0	12.9	23.5	48.0	100.7	212.7	450.1

8 To find the area of a pond situated near a straight road, perpendicular offsets at 20 m intervals were measured to the near and far boundaries of the pond. Use Simpson's rule with the following results to find the area of the pond.

Distance measured along road (m)	0	20	40	60	80	100	120
Length of offset to near boundary (m)		35	12	9	8	10	14	38
Length of offset to far boundary (m)		35	41	58	64	67	51	38

9 The following table gives the areas of cross-section of a tree trunk, measured every 1.5 m along its length. Estimate the volume of wood, using Simpson's rule.

Distance from base (m)	0.0	1.5	3.0	4.5	6.0	7.5	9.0
Area of cross-section (m²)	0.615	0.584	0.510	0.472	0.422	0.391	0.376

10 The areas at various heights of horizontal sections of a mound of gravel at a quarry are given below. Use Simpson's rule to determine the approximate volume and hence the mass of gravel in tonnes, if the density is 1.75 t/m^3.

Height from base (m)	0.0	1.0	2.0	3.0	4.0	5.0	6.0
Horizontal area of cross-section (m²)	50.5	38.4	32.1	24.6	15.2	7.1	0

11 If the correct answer is 68 sq. units, calculate as a percentage the error which arises when the area below the curve $y = x^3 + 1$ between the ordinates at $x = 0$ and $x = 4$ and above the x-axis is found using the mid-ordinate rule with four strips. Show that Simpson's rule is exact in this case.

12 Define the mean value of a waveform and find the mean value of the ramp waveform in fig. D2.16.

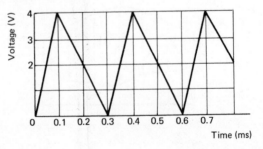

Fig. D2.16

134

13 What is the average value of a sinusoidal voltage with a maximum value of 240 V over (a) a half wave? (b) a full wave?

14 Find the mean value of the waveform in fig. D2.17.

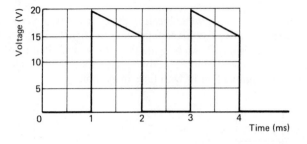

Fig. D2.17

D3 Centroids and moments of area

D3.1 Centroid of a lamina

Consider the irregular area in fig. D3.1 to be cut from a thin sheet (of say metal or plastic) with constant thickness and density. We call this a *uniform lamina*. On the lamina there will be a unique point G known as the *centroid*.

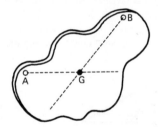

Fig. D3.1 Centroid of a lamina

The centroid of a lamina (like the centre of gravity of a solid object) is the point at which, for the purposes of calculation, we can consider all the mass to be located. In other words, if we were to make a small hole at G and pin the lamina loosely to a vertical board, the lamina would balance in any position we moved it to. The centroid of a lamina like the one in fig. D3.1 could be located experimentally by pinning the lamina loosely as described from a point A and allowing it to hang freely. It will come to rest with the centroid directly below A. After drawing a line vertically through A, the process is repeated choosing a second point B. The intersection of the two lines will be the point G.

135

To determine the position of the centroid mathematically for such an irregular shape as the one in fig. D3.1 is beyond the scope of this unit. However we can easily locate the centroid of many common shapes.

D3.2 Centroid of regular areas

A useful point to remember when locating the centroid of a plane area is that, when we can draw a *line of symmetry* through the figure, the centroid will lie somewhere along this line. If we imagine the area to be folded in two along such a line the two halves would exactly match with no overlapping. Figure D3.2 shows some simple figures with lines of symmetry.

Fig. D3.2 Lines of symmetry

If two lines of symmetry can be drawn on the area, the centroid will lie at their intersection. This is true for the rectangle, the circle, and the I-section (when both flanges have the same dimensions). To locate the centroid of a triangle we can consider it to be made up of a large number of approximate rectangles, as shown in fig. D3.3(a). The centroid of each narrow rectangle we know to be at its centre. The line joining all such centroids is a *median*, shown in fig. D3.3(a) drawn from B to the centre of AC. Other medians can be drawn from A and C and they intersect at the centroid of the triangle. Furthermore, it can be shown that the centroid is exactly one third of the way up each median from the side of the triangle which the median bisects.

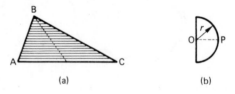

Fig. D3.3 Locating the centroid on a triangle and semicircle

In the case of the semicircle in fig. D3.3(b), we know that the centroid lies along OP, since this is a line of symmetry. We would expect (thinking in terms of a balance of areas) that it would lie nearer to O than to P. In fact it can be shown to lie $4r/3\pi$ (approximately $0.42r$) away from O. For example, a semicircle with radius 12 mm has its centroid 5.09 mm from the base along its line of symmetry.

Table D3.1 summarises the locations of centroids of some common shapes, which should be remembered.

Title	Figure	Centroid location
Rectangle	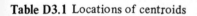	$\bar{x} = \dfrac{a}{2}$ $\bar{y} = \dfrac{b}{2}$
Triangle		$BE = \dfrac{BA}{2}$ and $EG = \dfrac{EC}{3}$ $CD = \dfrac{CA}{2}$ and $DG = \dfrac{DB}{3}$ $\bar{y} = h/3$
Isosceles triangle		$\bar{x} = \dfrac{BA}{2}$ $\bar{y} = \dfrac{CD}{3}$
Circle		$\bar{x} = r$ $\bar{y} = r$
Semicircle		$\bar{x} = r$ $\bar{y} = \dfrac{4r}{3\pi}$ or $0.42r$

Table D3.1 Locations of centroids

D3.3 First moment of area

Practical problems in bending and stress analysis require us to be able to locate the centroid in plane areas. Figure D3.4 shows an irregular lamina with

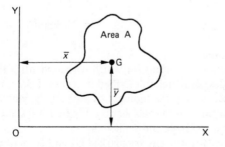

Fig. D3.4 Location of the centroid of a lamina

137

area A and centroid at G, which is a distance $\bar{y}$ measured perpendicularly from the OX axis and $\bar{x}$ from the OY axis.

We define the *first moment of area* about an axis as *the product of the area and the perpendicular distance of its centroid from the axis*. While we normally associate taking moments with the product of distance and force (e.g. weight), the mass of a thin lamina depends only on its area and, from our definition, we can consider the whole mass to be concentrated at the centroid. In all cases the axes must be in the plane of the area.

The first moment of area about the OY axis for the lamina in fig. D3.4 is therefore $A\bar{x}$, and about the OX axis $A\bar{y}$.

Example　Find the first moments of area of the rectangle in fig. D3.5 about both axes drawn.

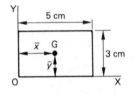

Fig. D3.5 Moment of area of a rectangle

The centroid of the rectangle is at G, where

$$\bar{x} = 2.5 \text{ cm} \quad \text{and} \quad \bar{y} = 1.5 \text{ cm}$$

$$\text{Area } A = 5 \times 3 = 15 \text{ cm}^2$$

First moment of area about OY axis $= A\bar{x}$

$$= 15 \times 2.5$$

$$= 37.5 \text{ cm}^3$$

First moment of area about OX axis $= A\bar{y}$

$$= 15 \times 1.5$$

$$= 22.5 \text{ cm}^3$$

D3.4 Composite areas

When an area consists of sections for which we can identify the centroid (rectangles, triangles, and circles described in section D3.2), the definition in section D3.3 is used to find the moment of area for the whole *composite area* by adding contributions from each section. Figure D3.6 shows a composite area comprised of a rectangle and semicircle.

Centroids of the two sections are marked G_1 and G_2, and the first moments area about axes OX and OY can be found as follows. A table is used to clarify

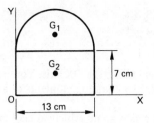

Fig. D3.6 First moment of area of a composite

Section	Area (cm²)	Distance of centroid from OY (cm)	Distance of centroid from OX (cm)	First moments of area (cm³)	
				About OY	About OX
	$\frac{1}{2} \times \pi \times (6.5)^2$ $= 66.37$	$\frac{13}{2}$ $= 6.50$	$7 + \frac{4 \times 6.5}{3 \times \pi}$ $= 9.76$	66.37×6.50 $= 431.4$	66.37×9.76 $= 647.8$
	13×7 $= 91$	$\frac{13}{2}$ $= 6.5$	$\frac{7}{2}$ $= 3.5$	91×6.5 $= 591.5$	91×3.5 $= 318.5$

Table D3.2

the method (Table D3.2). Adding values in the two columns on the right gives the total first moments of area for the composite:

first moment of area about OY = 431.4 + 591.5 = 1022.9 cm³

and   about OX = 647.8 + 318.5 = 966.3 cm³

D3.5 Location of the centroid

The method in section D3.1 of finding moments of area about two axes enables us to locate the centroid of composite shapes. This is done by equating the sum of moments of area of all sections about an axis to the product of total area and distance of the centroid from that axis, as follows.

The centroid of the composite area in fig. D3.6 will lie on the line of symmetry passing through G_1 and G_2. To locate it we need only consider moments about axis OX.

Let the centroid by $\bar{y}$ from the OX axis.

Total area of composite = 157.4 cm²

∴   total moment of area about OX = 157.4 × $\bar{y}$

Equating this to section components gives

157.4 × $\bar{y}$ = 647.8 + 318.5

139

$$\therefore \qquad \bar{y} = \frac{966.3}{157.4}$$

$$= 6.14 \text{ cm}$$

i.e. the centroid is 6.14 cm above the OX axis. You may like to prove that it is on the line of symmetry by equating moments of area about the OY axis.

By carefully choosing the axes about which to take moments, we can reduce the amount of arithmetic involved. If an axis passes through the centroid of any section of the composite area, the contribution of that section to the total moment of area about that axis will be zero. This follows from our definition of what a centroid is, in section D3.1. We can also make use of lines of symmetry as before.

Consider the T-shaped area in fig. D3.7; this consists of two rectangular areas. The figure is symmetrical about the vertical line YY', and so the centroid (shown as G) will be somewhere along it. We define a second axis XX' which is *orthogonal* (at right angles) to axis YY' and in the plane of the area. Table D3.3 shows how we equate the sum of moments of area for the sections to that of the composite area to find $\bar{y}$ (the distance of G from XX').

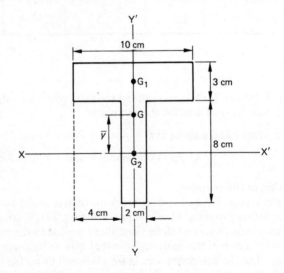

Fig. D3.7 Centroid of a composite area

Equating moments of area,

$$165 + 0 = 46\bar{y}$$

$$\therefore \qquad \bar{y} = 3.59 \text{ cm}$$

i.e. the centroid of the composite area is 3.59 cm above the XX' axis, or 7.59 cm from the base.

Section	Area (cm²)	Distance of centroid from axis XX' (cm)	First moment of area about XX' (cm³)
▭	10×3 $= 30$	$4 + 1.5$ $= 5.5$	30×5.5 $= 165$
▯	8×2 $= 16$	0	0
⊤	$30 + 16$ $= 46$	$\bar{y}$	$46\bar{y}$

Table D3.3

Note that, by placing the axis XX' through the known centroid (G_2) of one of the rectangles, we have eliminated the moment of area contribution from that rectangle, simplifying the calculation.

Example 1 Locate the centroid of the composite area shown in fig. D3.8.

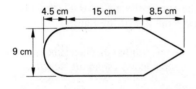

Fig. D3.8

Let XX' and YY' be orthogonal axes in the plane of the area and G_1, G_2, G_3 be the centroids of the three sections. The centroid of the complete area (G) will lie on the XX' axis (line of symmetry — see fig. D3.9). First moments of area are shown in Table D3.4.

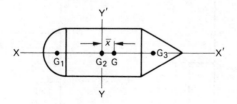

Fig. D3.9

Equating moments of area from the table,

$$205.06\bar{x} = 395.12 - 299.33$$

141

Section	Area (cm²)	Distance of centroid from axis YY' (cm)	First moment of area about YY' (cm³)
◖	31.81	-9.41 $(= 7.5 + \dfrac{4 \times 4.5}{3\pi})$	-299.3
▭	135	0	0
▷	38.25	10.33 $(= 7.5 + \dfrac{8.5}{3})$	395.1
◖▷	205.06	$\bar{x}$	$205.06\,\bar{x}$

Table D3.4

$$\therefore \qquad \bar{x} = \frac{95.79}{205.06}$$
$$= 0.47 \text{ cm}$$

i.e. the centroid of the composite area is 0.47 cm to the right of G_2, or 12.47 cm from the extreme left of the figure.

Note that (as when drawing graphs) distances measured to the left of or below the origin of our orthogonal axes system are *negative*.

Example 2   Locate the centroid of the plane area in fig. D3.10(a).

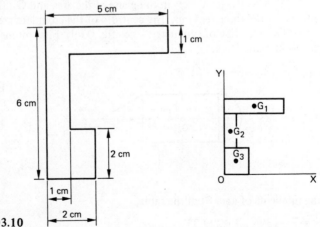

Fig. D3.10

Let OX and OY be orthogonal axes and G_1, G_2, G_3 be the centroids of three rectangular sections, with the centroid of the composite area at $(\bar{x}, \bar{y})$; see fig. D3.10(b).

Section	Area (cm²)	Distance of centroid from OY (cm)	Distance of centroid from OX (cm)	Moment of area about OY (cm³)	Moment of area about OX (cm³)
▭	5	2.5	5.5	12.5	27.5
▯	3	0.5	3.5	1.5	10.5
▢	4	1	1	4	4
L	12	$\bar{x}$	$\bar{y}$	$12\bar{x}$	$12\bar{y}$

Table D3.5

Equating moments of area about OY (Table D3.5),

$$12\bar{x} = 12.5 + 1.5 + 4$$

$$\therefore \quad \bar{x} = \frac{18}{12} = 1.5 \text{ cm}$$

and about OX,

$$12\bar{y} = 27.5 + 10.5 + 4$$

$$\therefore \quad \bar{y} = \frac{42}{12} = 3.5 \text{ cm}$$

i.e. the centroid is located 1.5 cm from the left edge and 3.5 cm from the base of the figure.

Note that in this example the centroid lies *outside* the material of the section.

Exercise D3
1 Define the terms (a) 'lamina', (b) 'centroid'.
2 Determine $\bar{x}$ and $\bar{y}$ in each of the laminae in fig. D3.11, where G is the centroid.

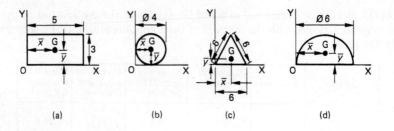

(a) (b) (c) (d)

Fig. D3.11

3 Define the term 'first moment of area'.
4 Calculate the first moments of area about axes OX and OY for the figures in question 2.
5 Determine the first moments of area about axes OX and OY for the symmetrical areas in fig. D3.12.

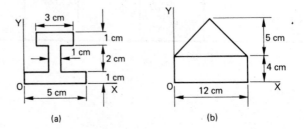

(a) (b)

Fig. D3.12

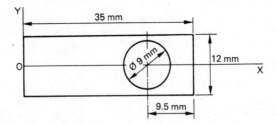

Fig. D3.13

6 By subtracting for the circular area removed, calculate the first moment of area about the OY axis of the area in fig. D3.13.
7 Calculate how far from the OY axis the centroid of the area in question 6 lies.
8 Determine the distance from the base to the centroid of the rivet cross-section in fig. D3.14.

144

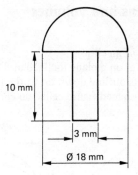

Fig. D3.14

9 Calculate the height of the centroid from the base of the simplified rail cross-section in fig. D3.15.

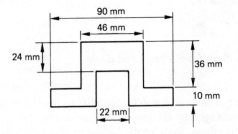

Fig. D3.15

10 Explain why no calculation of moments of area is necessary to locate the horizontal position of the centroid for the section in fig. D3.15.
11 Calculate the distance of the centroid from axes OX and OY for the subframe cross-section in fig. D3.16.

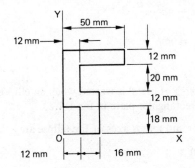

Fig. D3.16

D4 Curvilinear areas and volumes

D4.1 Area and perimeter of an ellipse

If we draw a circle, radius a, on a plane inclined at $\theta°$ to the horizontal (see fig. D4.1(a)) and project its outline down on to the horizontal plane, we obtain a symmetrical oval-shaped figure called an *ellipse*.

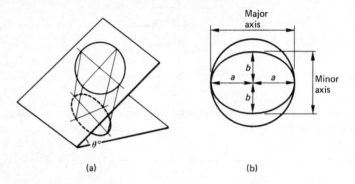

Fig. D4.1 The ellipse from a projected circle

The diameter of the circle which is parallel to the line of intersection of the planes is unchanged by the projection (see section D6.1) and becomes the *major axis* of the ellipse, with length $2a$ (see fig. D4.1(b)). At right angles to the major axis is the *minor axis*, length $2b$, where $2b = 2a \cos \theta$.

The area of the ellipse is found as follows:

$$\text{area of ellipse} = \text{area of circle} \times \cos \theta$$

$$= \pi a^2 \times \cos \theta$$

but

$$b = a \cos \theta$$

$\therefore$

$$\text{area} = \pi a^2 \times \frac{b}{a}$$

$$= \pi a b$$

Example 1 What area of metal is wasted when an ellipse is cut from the 80 mm by 120 mm rectangular plate in fig. D4.2.

Lengths of the major and minor axes of the ellipse are

$$2a = 120 \text{ mm}$$

$$2b = 80 \text{ mm}$$

146

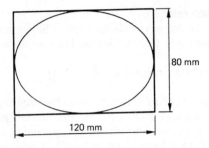

Fig. D4.2

∴   area of ellipse $= \pi ab = \pi \times 60 \times 40$

$$= 7540 \, \text{mm}^2$$

area of metal wasted $= (120 \times 80) - 7540$

$$= 2060 \, \text{mm}^2$$

The perimeter of an ellipse $= \pi(a + b)$.

Note that, as b gets closer to a, the perimeter approaches $2\pi a$, which is the circumference of a circle radius a.

Example 2   A 150 mm diameter pipe is cut at an angle of $40°$ to its length and welded to a flat plate. Find the length of the weld.

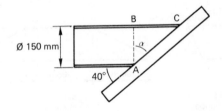

Fig. D4.3

Figure D4.3 shows the pipe and plate. The line of contact between the pipe and the plate is an ellipse with major-axis length AC and minor-axis length 150 mm.

Angle $\alpha = (90 - 40)° = 50°$

∴ $AC = \dfrac{AB}{\cos \alpha} = \dfrac{150}{\cos 50°} = 233.4 \, \text{mm}$

Perimeter of the ellipse $= \pi(a + b)$

$$= \pi \left(\frac{233.4}{2} + \frac{150}{2} \right)$$

∴   length of weld $= 602.2 \, \text{mm}$

147

D4.2 The prismoidal rule

This rule may be regarded as a modified form of Simpson's rule (see sections D2.4 and D2.5) using only three ordinates. The formula may be stated as

$$\text{volume} = \frac{h}{6}(\text{base} + \text{top} + 4 \times \text{mid-section})$$

where h is the perpendicular distance between base and top, and the base, top, and mid-section are all parallel planes. This formula gives exact values for the volumes of many objects, including cones, pyramids, and spheres. It is also exact in the case of the volume of any *prismoid*. A prismoid is a solid with parallel plane ends and all its other sides trapeziums.

Example A section of a hill which is to be removed during construction of a motorway is a prismoid. Use the dimensions in fig. D4.4 to find the volume of earth to be excavated.

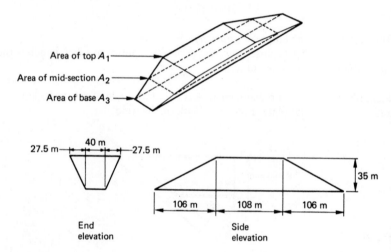

Fig. D4.4

By the prismoidal rule,

$$\text{volume} = \frac{h}{6}(A_1 + A_3 + 4A_2)$$

where h = depth = 35 m

A_1 = area of base = 320 x 40 = 12 800 m²

A_3 = area of top = 108 x 95 = 10 260 m²

A_2 = area of the mid-section

148

The mid-section is a rectangle with sides

$$\frac{320 + 108}{2} = 214 \text{ m} \quad \text{and} \quad \frac{40 + 95}{2} = 67.5 \text{ m}$$

$\therefore \qquad A_2 = 214 \times 67.5 = 14\,445 \text{ m}^2$

giving

$$\text{volume} = \frac{35}{6}\,[12\,800 + 10\,260 + (4 \times 14\,445)]$$

$$= 4.72 \times 10^5 \text{ m}^3$$

D4.3 The theorem of Pappus for the volume of a solid

The theorem of Pappus states that, when a plane area revolves about an axis which is in the same plane but does not cut the area, the volume generated is equal to the product of the area and the distance moved by the centroid of the area;

i.e. $\quad \text{volume} = 2\pi\bar{y}A$

where $\qquad \bar{y} = $ distance of the centroid from the axis

and $\qquad A = $ area

This theorem is used both in finding volumes and in locating centroids when the volume of revolution is known.

Example 1 Find the volume of the ring generated when the circle in fig. D4.5, radius 10 mm, revolves through 360° about an axis 25 mm from the centre of the circle in the same plane.

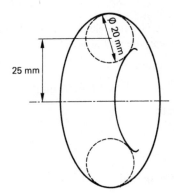

Fig. D4.5

The centroid of the circle is at its centre, and the area is $\pi \times 10^2$ mm². By the theorem of Pappus,

$$\text{volume} = 2\pi\bar{y}A$$

where $\bar{y}$ = distance of centroid from axis

$\qquad$ = 25 mm

and $\qquad A = \pi \times 10^2 = 314.2 \, \text{mm}^2$

$\therefore \quad$ volume $= 2\pi \times 25 \times 314.2$

$\qquad\qquad\qquad = 4.94 \times 10^4 \, \text{mm}^3$

Note that if the solid ring were cut and straightened out to form a cylinder, the length would be equal to $2\pi\bar{y}$ and we would obtain the same volume using the formula for a cylinder.

Example 2 A lathe is used to make a symmetrical triangular groove in the end of a 50 mm diameter aluminium bar with the dimensions in fig. D4.6. Calculate the volume of metal removed.

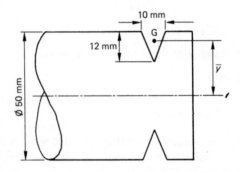

Fig. D4.6

The centroid G of the isosceles triangle lies one third of the way along a vertical line passing through the mid-point of the base (see section D3.2);

$\therefore \quad \bar{y} = 25 - 12/3 = 21 \, \text{mm}$

The area of the triangle $= \frac{1}{2} \times 10 \times 12 = 60 \, \text{mm}^2$

By the theorem of Pappus,

$\qquad$ volume $= 2\pi\bar{y}A$

$\qquad\qquad\qquad = 2\pi \times 21 \times 60$

i.e. the volume of metal removed $= 7917 \, \text{mm}^3$.

Example 3 Use the theorem of Pappus to prove that the centroid of a semi-circular area, radius r, lies at a distance $4r/3\pi$ from its base.

Figure D4.7 shows a semicircular area, radius r, with its centroid G a distance $\bar{y}$ from the base. If we revolve the area through $360°$ about the base, the volume generated will be a sphere radius r and volume $4\pi r^3/3$.

150

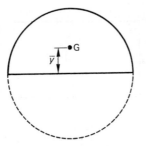

Fig. D4.7

From the theorem of Pappus,

$$\text{volume} = 2\pi \bar{y} A$$

$$\therefore \qquad \frac{4\pi r^3}{3} = 2\pi \bar{y} \times \frac{\pi r^2}{2}$$

or $\qquad \bar{y} = \frac{4\pi r^3}{3\pi^2 r^2} = \frac{4r}{3\pi}$

Exercise D4

1 Calculate the area and perimeter of an ellipse with major- and minor-axis lengths 45 mm and 28 mm respectively.

2 A path 2 m wide is laid around an elliptical area of grass with major axis 20 m and minor axis 11 m. Find the area of the path.

3 A circular steel bar 50 mm diameter is cut at an angle so that the area of the elliptical section is 2118 mm². Find the length of the major axis of the cut surface and the angle it makes with the length of the bar.

4 An elliptical hole is to be cut in a roof which slopes at 30° to the horizontal to take a vertical 150 mm diameter pipe. Calculate the area of the hole.

5 Use the prismoidal rule to find the volume of the convergent–divergent nozzle in fig. D4.8.

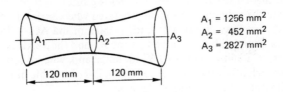

$A_1 = 1256$ mm²
$A_2 = 452$ mm²
$A_3 = 2827$ mm²

Fig. D4.8

6 A steel barrel stands 850 mm high and the diameter at both top and bottom is 600 mm. The diameter mid-way between the ends is 780 m. Use the prismoidal rule to estimate the capacity of the barrel in litres.

7 State the theorem of Pappus for the volume of a solid and use it to determine the volume generated when the rectangular area ABCD in fig. D4.9

151

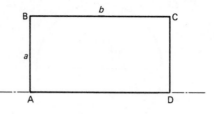

Fig. D4.9

revolves through 360° about the line AD. Show that your answer is the volume of a cylinder.

8 Use the theorem of Pappus to calculate the volume of a rubber moulding with the cross-section shown in fig. D4.10, if the moulding completely surrounds a circular window 450 mm in diameter and its area of cross-section is 750 mm².

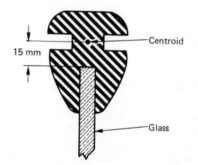

15 mm

Centroid

Glass

Fig. D4.10

9 Mercury is poured into a U-tube made from 8 mm bore glass tube until it forms a semicircle with internal radius 30 mm. Calculate the mass of mercury in the tube if the density of mercury is 13 600 kg/m³.
10 Calculate the volume of the solid generated when a triangle ABC revolves about side AC through 120°, if AB = BC = 50 mm and AC = 38 mm.
11 The area bounded by the curve $y = x^2 + 1$ and the x-axis between ordinates at $x = 1$ and $x = 3$ is 10.67 sq. units; and, when it revolves about the x-axis through 360°, the volume generated is 212.8 cubic units. Calculate the y-co-ordinate of the centroid of this area.

D5 Basic trigonometry

D5.1 Fundamental definitions and relationships
Consider carefully the right-angled triangle ABC shown in fig. D5.1. Although the letters A, B, and C are put there to label the points where the sides of the

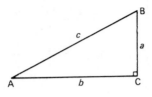

Fig. D5.1

triangle intersect (called *vertices*), the angles of the triangle are also often referred to by letters. Thus 'angle A', or '$\angle A$', may be used to denote the angle at vertex A. If this is likely to cause confusion, for example if there is more than one angle at a particular point, then it is better either to refer to it as angle BAC, i.e. the angle between BA and AC, or else to place a small letter such as x, y, z, α, β, θ, or ϕ in the angle so that the particular angle referred to may be easily identified.

In any right-angled triangle, the side opposite to the right-angle is called the *hypotenuse*, the other two sides being called the *opposite* and the *adjacent*, according to their position relative to the angle under consideration. Thus, in fig. D5.1, relative to angle A the side BC is opposite and AC is adjacent, whereas the reverse is true relative to angle B. It is usual to denote the side opposite angle A as a, the side opposite angle B as b, etc.

The three most important trigonometrical ratios are defined as follows:

the *sine* of angle A: $\qquad \sin A = \dfrac{\text{opposite}}{\text{hypotenuse}}$

the *cosine* of angle A: $\qquad \cos A = \dfrac{\text{adjacent}}{\text{hypotenuse}}$

the *tangent* of angle A: $\qquad \tan A = \dfrac{\text{opposite}}{\text{adjacent}}$

The reciprocals of these are also very useful, and these are as follows:

the *cosecant* of angle A: $\qquad \operatorname{cosec} A = \dfrac{1}{\sin A}$

the *secant* of angle A: $\qquad \sec A = \dfrac{1}{\cos A}$

the *cotangent* of angle A: $\qquad \cot A = \dfrac{1}{\tan A}$

Let us now state these ratios for angles A and B of fig. D5.1 in terms of the sides of the triangle:

$$\sin A = \frac{\text{BC}}{\text{AB}} = \frac{a}{c} \qquad\qquad \sin B = \frac{\text{AC}}{\text{AB}} = \frac{b}{c}$$

$$\cos A = \frac{AC}{AB} = \frac{b}{c} \qquad \cos B = \frac{BC}{AB} = \frac{a}{c}$$

$$\tan A = \frac{BC}{AC} = \frac{a}{b} \qquad \tan B = \frac{AC}{BC} = \frac{b}{a}$$

$$\operatorname{cosec} A = \frac{AB}{BC} = \frac{c}{a} \qquad \operatorname{cosec} B = \frac{AB}{AC} = \frac{c}{b}$$

$$\sec A = \frac{AB}{AC} = \frac{c}{b} \qquad \sec B = \frac{AB}{BC} = \frac{c}{a}$$

$$\cot A = \frac{AC}{BC} = \frac{b}{a} \qquad \cot B = \frac{BC}{AC} = \frac{a}{b}$$

As the angles of a triangle add up to 180°, and angle C is a right angle, it follows that $A + B = 90°$. Angles A and B are said to be *complementary* angles.

The following are some of the more important relationships arising from the above:

$$\tan A = \frac{\sin A}{\cos A} \qquad\qquad \cot A = \frac{\cos A}{\sin A}$$

$$\sin A = \cos (90 - A) \qquad \cos A = \sin (90 - A)$$

$$\tan A = \cot (90 - A) \qquad \cot A = \tan (90 - A)$$

D5.2 Angles of any magnitude

So far we have only considered acute angles, i.e. angles less than 90°, but we must also be able to find the trigonometrical ratios of obtuse, reflex, and sometimes even negative angles.

Consider a circle of unit radius as shown in fig. D5.2. Since OP, the hypotenuse of triangle OAP, is of unit length, its projections on the horizontal and vertical axes through O are simply $\cos x$ and $\sin x$ respectively. Thus, for any angle x, the length OA on the horizontal axis will give the cosine of the angle, and the length OB on the vertical axis will give the sine of the angle.

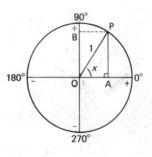

Fig. D5.2

154

Consider the position of OP when the angle x is obtuse. Then, for an angle between 90° and 180°, the triangle will be in the second quadrant of the circle, so, although the projection for sin x will be on the positive vertical axis, the projection for cosine will be on the *negative* horizontal axis, thus showing that cosines of all obtuse angles are negative.

Further anticlockwise rotation of OP will give reflex angles, and it can be seen that in the third quadrant (180° to 270°) both sine and cosine will be negative, while in the fourth quadrant (270° to 360°) the sines are negative and the cosines positive.

Negative angles are given by rotating clockwise from the zero position, and angles greater than 360° are given by rotating a complete revolution plus the amount by which the angle exceeds 360°.

For tangents, consider fig. D5.3: tan x in triangle OQR is QR/OR, but OR = 1 for the circle of unit radius; therefore tan x is represented by the length of the intercept QR on the tangent to the circle at R. For angles in the first and third quadrants the tangents are positive, whereas for angles in the second and fourth quadrants the tangents are negative.

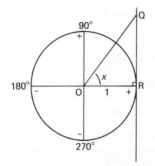

Fig. D5.3

Most scientific calculators will give values of sin x, cos x, or tan x with the correct sign for any angle x.

To find the value of the sine, cosine, or tangent of any angle greater than 90° using tables, first fix its sign as above, then refer to the angle symmetrically placed in the first quadrant and look up the value of this in the appropriate table.

Figure D5.4 indicates which trigonometrical ratios are positive in each quadrant.

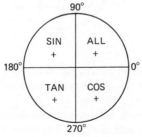

Fig. D5.4

155

Example 1   Find the sine, cosine, and tangent of $140°$.

From the composite diagram of fig. D5.5, we see that the sine is positive, while the cosine and tangent are negative. The corresponding angle in the first quadrant is $180° - 140° = 40°$,

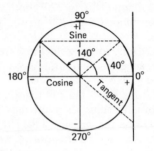

Fig. D5.5

$\therefore \quad \sin 140° = \sin 40° = 0.6428$

$\cos 140° = -\cos 40° = -0.7660$

$\tan 140° = -\tan 40° = -0.8391$

Check that your calculator gives the above values directly.

Example 2   Find the angles between $0°$ and $360°$ which have a sine of -0.6.

Referring to a table of natural sines, we see that $\sin 36°52' = +0.6$. From fig. D5.6 we see that the two angles which have a sine of -0.6 are $(180° + 36°52')$ and $(360° - 36°52')$, i.e. $216°52'$ and $323°8'$.

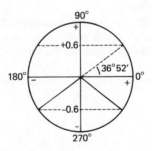

Fig. D5.6

Using a calculator instead of four-figure tables would yield $36.87°$ instead of $36°52'$, and hence the two angles which have a sine of -0.6 are $216.87°$ and $323.13°$.

The appropriate function button on a calculator may be labelled either 'arcsin' or 'sin⁻¹', both of which are used to mean 'the angle which has a sine

156

value of'. Using this notation we could have phrased the question in an alternative form asking for values of arcsin (-0.6) between $0°$ and $360°$.

Example 3   Find the values of (a) cosec $200°$, (b) sec $320°$, (c) cot $330°$.

a) $\operatorname{cosec} 200° = \dfrac{1}{\sin 200°} = \dfrac{1}{-\sin 20°} = -\dfrac{1}{0.3420} = -2.924$

b) $\sec 320° = \dfrac{1}{\cos 320°} = \dfrac{1}{\cos 40°} = \dfrac{1}{0.7660} = 1.305$

c) $\cot 330° = \dfrac{1}{\tan 330°} = \dfrac{1}{-\tan 30°} = \dfrac{1}{-0.5773} = 1.732$

With most calculators, these values can be obtained more easily. For example, to find cosec $200°$ we simply enter 200 with the calculator set to work in degrees and use the sine function, to give -0.3420, followed by the reciprocal to give -2.924.

D5.3 Radians
Angles can be measured either in degrees, minutes and seconds or, alternatively, in radians and milliradians. The relationship between the units is that π radians are equivalent to $180°$,

i.e.   1 radian $= \dfrac{180°}{\pi}$, or approximately $57°18'$

A complete revolution of $360°$ is 2π radians. Since the circumference of a circle is 2π x radius, we can define a radian as the angle subtended at the centre of a circle by an arc equal in length to the radius.

Degrees and radians
To convert degrees to radians we multiply by π and divide by $180°$.

Example 1   Convert to radians (a) $45°$, (b) $30°$.

Taking π as 3.1416,

a) $45° \times \dfrac{3.1416}{180°} = \dfrac{3.1416}{4} = 0.7854$ rad

b) $30° \times \dfrac{3.1416}{180°} = \dfrac{3.1416}{6} = 0.5236$ rad

Conversion is easier if an electronic calculator is available which will enter an accurate value for π at the press of a button. Alternatively, many books of four-figure tables include a table for the conversion of degrees to radians.
To convert radians to degrees we multiply by $180°$ and divide by π.

Example 2   Convert 3.927 radians to degrees.

Taking π as 3.1416,

$$3.927 \text{ radians} = 3.927 \times \frac{180°}{3.1416} = 225°$$

For small angles, measurements are made in milliradians (1 rad = 1000 mrad). Conversions can be made using the following relationships:

$$1° = 17.45 \text{ mrad} \quad \text{and} \quad 1 \text{ mrad} = 3'26''$$

D5.4 Angular rotation

A straight strip of tape stuck on the face of a gear wheel from axis to perimeter will enable us to see that, in one revolution of a gear wheel, an individual tooth is rotated through an angle of 360°. Since 360° = 2π radians, we see that one revolution is 2π radians. In radian measure, therefore, all even multiples of π correspond to complete revolutions, e.g. 4π radians will be 2 revolutions, 6π radians will be 3 revolutions, etc.

If a shaft or a pulley is rotating at 3 revolutions per second, then the angular rotation must be $3 \times 2\pi$ radians per second. In general terms, a rotation of n revs per second will give an angular velocity of $2\pi n$ radians per second. This is obviously true whether n is a whole number or a fraction.

Example A shaft is rotating at 50 rev/min. Express this in radians per second.

50 rev/min is 50 revolutions per minute

which is $\dfrac{50}{60}$ revolutions per second

which is $\dfrac{50}{60} \times 2\pi$ radians per second

i.e. $\dfrac{5\pi}{3}$ radians per second

which is 5.236 radians per second.

D5.5 Arcs and angles

From the circumference formula $c = 2\pi r$, we can deduce the arc length corresponding to any angle in radians. Since a circumference length of $2\pi r$ corresponds to one complete revolution of 2π radians, an angle of 1 radian will give a circular arc of length equal to the radius. For a centre angle of θ radians the arc length s at radius r is given by $r\theta$. If the radius is in millimetres, the arc length will also be in millimetres for, in the formula $s = r\theta$, s and r must be in the same units.

Example 1 A pendulum of length 2.0 m swings through an angle of 0.15 radians in a single swing. Find the length of the arc traced by the pendulum bob.

$$s = r\theta \quad \text{where } r = 2.0\,\text{m} \quad \text{and} \quad \theta = 0.15\,\text{rad}$$

$$\therefore \quad s = 2 \times 0.15$$

$$= 0.30\,\text{m}$$

We can transpose the formula $s = r\theta$ to give $\theta = s/r$, from which we can determine the centre angle corresponding to a given arc length.

Example 2   Find the angle of lap if 210 mm of a belt drive are in contact with a pulley of radius 150 mm.

$$\theta = \frac{s}{r} \quad \text{where } s = 210\,\text{mm} \quad \text{and} \quad r = 150\,\text{mm}$$

$$\therefore \quad \theta = \frac{210}{150} = \frac{21}{15} = \frac{7}{5} = 1.4\,\text{radians}$$

The principle of angular rotation combined with the formula for the length of an arc enables us to solve problems concerning gear trains and belt drives.

Example 3   A gear wheel of 200 mm radius is being driven by a second gear wheel of radius 160 mm. Find the angle turned through by the first gear wheel for each revolution of the second.

For the 160 mm gear wheel, $c = 2\pi r = 320\pi$ mm and the 200 mm gear wheel moves through an arc of equal length.

$$\theta = \frac{s}{r} = \frac{320\pi}{200}$$

$$= 1.6\pi\,\text{radians}$$

$$= 288°$$

D5.6 Area of sector

The area of a complete circle is πr^2 and this corresponds to a complete revolution of 360° or 2π radians. By proportion, a sector with an angle of θ radians (fig. D5.7) must have an area of

$$\frac{\theta}{2\pi} \times \pi r^2 = \tfrac{1}{2}r^2\theta$$

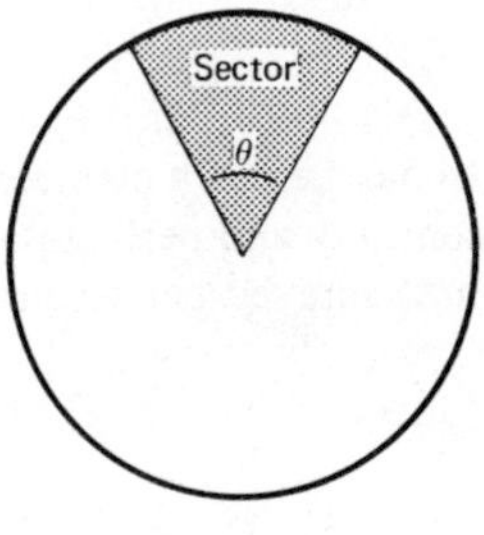

Fig. D5.7

Example   A floodlight can spread its illumination over an angle of 0.75 radians to a distance of 40 metres. Find the area thus floodlit.

$$A = \tfrac{1}{2}r^2\theta = \tfrac{1}{2} \times 40^2 \times 0.75 = 600\,\text{m}^2$$

Exercise D5

1   For each of the following angles find the numerical values of sine, cosine, tangent, cosecant, secant, and cotangent: (a) 62.5°, (b) 100°12′, (c) 212.25°, (d) 335°36′.

2   Evaluate (a) arcsin 0.753, (b) arccos 0.447, (c) arctan 2.500.

3   If $x = 152.5°$, find the value of $4 \sin x - 3 \cos x$.

4   When $x = 0.65$ radians, find $\sin x$ and $\sec x$.

5   Find the value of $\operatorname{cosec} x - \sin 2x$ when $x = 104°$.

6   If $\theta = 1.43$ rad, find (a) $\sin x$, (b) $\sin^2 x$, (c) $\operatorname{cosec} x$.

7   Find the cosine and cosecant of the following angles: (a) 4 rad, (b) 1.8π rad, (c) 150 mrad.

8   Find the length of arc which subtends an angle of 72° at the centre of a circle radius 10.5 m.

9   A circle has a diameter of 240 mm. Find the angle at the centre subtended by an arc of length 150 m.

10   Show that if l is the length of the arc of a sector of a circle radius r, then the area of the sector is $\tfrac{1}{2}rl$. Hence find the area of a sector radius 5 m and arc length 9 m. Find also the angle of the sector.

11   An automatic spray system used in horticulture sends out a spray to a distance of 1.6 m over an angle of 1.5 radians in each of two directions simultaneously. Find the area covered by the spray.

12   A rubber strip used as a seal has a uniform cross-section as shown in fig. D5.8. The curved top is an arc of a circle of length 40 mm. The curved base is an arc of a circle of length 36 mm. The thickness of the strip is a uniform 5 mm. Calculate the area of this cross-section.

Fig. D5.8

13   A spray boom of length 5 m lies on a diameter across a circular sewage tank. The boom is pivoted centrally and performs one complete revolution every 13 seconds. Find the area sprayed per second.

D6 Applications of trigonometry

D6.1 Lengths and areas on an inclined plane

Lengths
We can relate lengths on an inclined plane to corresponding lengths on a plan by using the trigonometrical ratios defined in section D5.1.

Figure D6.1 shows a rectangular plane ABEF inclined at an angle θ to a horizontal rectangular plane ABCD. We can see from the right-angled triangle BEC that BC = BE cos θ and from triangle AFD that AD = AF cos θ. The lines BE and AF are called *lines of greatest slope* – they are on the inclined plane and are perpendicular to AB, the line of intersection of the two planes.

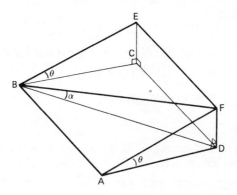

Fig. D6.1

Any line of greatest slope on the inclined plane will be longer than its projection on to the horizontal plane by a factor $1/\cos \theta$. Lines on the inclined plane which are parallel to AB will be the same length as their projections on to the horizontal plane. For example, EF is obviously the same length as CD, since EFDC is a rectangle.

Lines on the plane ABEF which are not lines of greatest slope will be longer than their projections on to plane ABCD by a factor lying between zero and $1/\cos \theta$, depending on the angle between the line itself and the line of intersection of the planes. For example, the angle α made between the line BF and plane ABCD is smaller than angle θ. The lengths BD and BF are related by the equation BD = BF cos α.

Example Two vertical pipes shown in fig. D6.2 pass through a horizontal ceiling and a roof which is inclined at $28°$ to the ceiling. If the pipe centres lie along a line which is perpendicular to the line of intersection of the roof and ceiling, and the distance between the centres is 425 mm, measured at the ceiling, what will be the corresponding measurement made along the roof?

161

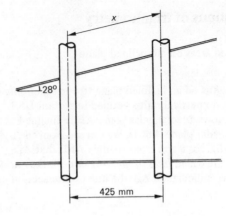

Fig. D6.2

The pipe centres lie along a line of greatest slope and so the distance between them measured along the roof is given by

$$x = \frac{425}{\cos 28°} = 481.3 \text{ mm}$$

Areas
In fig. D6.1, the area of rectangle ABCD is AB × BC and the area of ABEF is AB × BE. We have already seen that BC = BE cos θ, and so

area of ABCD = AB × BE cos θ

= area of ABEF × cos θ

Areas measured on the inclined plane will be larger than their projections on to the horizontal plane by a factor $1/\cos \theta$.

Example The flow of air through a rectangular air-conditioning duct shown in fig. D6.3 is controlled by the rotation in the duct of a rectangular plate with the same dimensions. Find the angle x which will reduce the effective area of the duct to one third of its maximum.

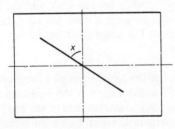
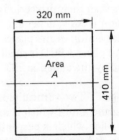

Fig. D6.3

162

The maximum area of the duct $= 320 \times 410 = 1.312 \times 10^5 \, \text{mm}^2$

When the plate makes an angle x with an axis perpendicular to the length of the duct,

area $A = (1.312 \times 10^5) \times \cos x$

When the effective area of the duct is reduced to one third of the maximum,

area $A = (1.312 \times 10^5) \times \frac{2}{3}$

$\therefore \qquad \cos x = \frac{2}{3}$

$\qquad x = 48.19°$

D6.2 Area of a triangle

Considering again a triangle ABC, right-angled at C as in fig. D6.4, the adjacent side AC represents a plan view of the hypotenuse AB. It gives the *projection* of AB on to the horizontal plane. Similarly, BC is the projection of AB on to the vertical plane.

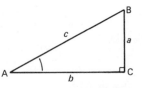

Fig. D6.4

Now $\dfrac{AC}{AB} = \cos A$ and $\dfrac{BC}{AB} = \sin A$

$\therefore \quad AC = AB \cos A$ and $BC = AB \sin A$

or $\quad b = c \cos A$ and $a = c \sin A$

We know that the area of a triangle is given by $\frac{1}{2}$ base x height, but it is not always possible to measure the height directly. One alternative is to measure two sides of the triangle and the angle between them (see fig. D6.5).

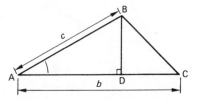

Fig. D6.5

Considering BD as the vertical projection of AB, we have the height of the triangle, BD, given by AB sin A, i.e. c sin A. Multiplying this by the base length b and halving it for the area of the triangle, we have

163

$$\text{area} = \tfrac{1}{2}bc \sin A$$

and similarly the alternative forms

$$\text{area} = \tfrac{1}{2}ab \sin C$$

$$\text{area} = \tfrac{1}{2}ac \sin B$$

Example 1   Find the area of triangle ABC if AB = 35.2 mm, AC = 27.7 mm and angle $A = 61.8°$.

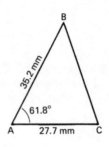

Fig. D6.6

From fig. D6.6,

$$\text{area} = \tfrac{1}{2}bc \sin A$$

$$= \tfrac{1}{2} \times 35.2 \times 27.7 \times \sin 61.8°$$

$$= 429.7 \text{ mm}^2$$

Example 2   Determine the area of a building site which forms a quadrilateral PQRS if PQ = 35.1 m, QR = 22.6 m, RS = 28.8 m, SP = 12.8 m, $\angle P = 61°$, and $\angle R = 73°$.

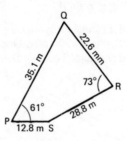

Fig. D6.7

From fig. D6.7,

$$\text{area of site} = \text{area of triangle PQS} + \text{area of triangle QRS}$$

$$= \tfrac{1}{2} \times 35.1 \times 12.8 \times \sin 61° + \tfrac{1}{2} \times 22.6 \times 28.8 \times \sin 73°$$

$$= 507.7 \text{ m}^2$$

164

An alternative formula for finding the area of a triangle is given in section D6.5.

D6.3 The sine rule

The area of a triangle is given by any of the three forms obtained above:

$$\tfrac{1}{2}bc \sin A = \tfrac{1}{2}ac \sin B = \tfrac{1}{2}ab \sin C$$

From this, if we divide through by $\tfrac{1}{2}abc$,

$$\frac{\sin A}{a} = \frac{\sin B}{b} = \frac{\sin C}{c}$$

This relationship is known as the *sine rule*, and can be used either in the above form or in the inverted form:

$$\frac{a}{\sin A} = \frac{b}{\sin B} = \frac{c}{\sin C}$$

Alternatively, the sine rule may be derived as follows (fig. D6.8).

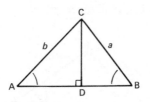

Fig. D6.8

From triangle ACD, $\quad CD = b \sin A$
From triangle BCD, $\quad CD = a \sin B$

$\therefore \qquad\qquad\qquad a \sin B = b \sin A$

i.e. $\qquad\qquad\qquad \dfrac{a}{\sin A} = \dfrac{b}{\sin B}$

By dropping a perpendicular from A on to BC and applying the same method,

$$\frac{b}{\sin B} = \frac{c}{\sin C}$$

and thus,

$$\frac{a}{\sin A} = \frac{b}{\sin B} = \frac{c}{\sin C}$$

The sine rule may be used to find another angle when given two sides and an angle opposite to one of them, or to find another side when given one side and two angles.

Example 1   In a triangle ABC, $a = 4$, $b = 5$, and angle $A = 52°$. Find angle B.

From the sine rule,

$$\frac{a}{\sin A} = \frac{b}{\sin B}$$

i.e. $$\frac{4}{\sin 52°} = \frac{5}{\sin B}$$

$$\sin B = \frac{5 \sin 52°}{4}$$

$$= \frac{5 \times 0.7880}{4}$$

$$= 0.9850$$

$\therefore$ $B = 80°4'$

Example 2   A workshop 8 m wide has a span roof which slopes at $35°$ on one side and $42°$ on the other. Find the length of the roof slopes.

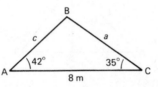

Fig. D6.9

Figure D6.9 shows a section of the roof. By subtraction, the angle at the ridge must be $103°$.

Applying the sine rule,

$$\frac{a}{\sin 42°} = \frac{8}{\sin 103°}$$

$\therefore$ $$a = \frac{8 \sin 42°}{\sin 103°}$$

$$= 5.49 \, m$$

Similarly,

$$\frac{c}{\sin 35°} = \frac{8}{\sin 103°}$$

$\therefore$ $$c = \frac{8 \sin 35°}{\sin 103°}$$

$$= 4.71 \, m$$

i.e. the two roof slopes are 5.49 m and 4.71 m (correct to the nearest 10 mm).

166

D6.4 The cosine rule

In fig. D6.10, BD is the perpendicular from B on to AC. As AD $= x$, DC $= b - x$.

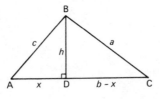

Fig. D6.10

Applying the theorem of Pythagoras to each of the triangles ABD and BCD,

$$a^2 = h^2 + (b - x)^2$$

i.e. $\quad a^2 = h^2 + b^2 - 2bx + x^2$

and $\quad c^2 = h^2 + x^2$

Subtracting,

$$a^2 - c^2 = b^2 - 2bx$$

but $\quad\quad x = c \cos A$

$\therefore \quad a^2 - c^2 = b^2 - 2bc \cos A$

i.e. $\quad\quad a^2 = b^2 + c^2 - 2bc \cos A$

Similarly,

$$b^2 = a^2 + c^2 - 2ac \cos B$$

and $\quad\quad c^2 = a^2 + b^2 - 2ab \cos C$

These forms are used to find the third side of a triangle when given two sides and the angle between them.

If all three sides are known and it is required to find an angle, then the cosine rule is more conveniently expressed in one of the following forms:

$$\cos A = \frac{b^2 + c^2 - a^2}{2bc}$$

$$\cos B = \frac{a^2 + c^2 - b^2}{2ac}$$

$$\cos C = \frac{a^2 + b^2 - c^2}{2ab}$$

Example 1   Because of obstructions, it is not possible to measure directly the distance between two points A and B. Measurements are therefore taken from another point C, as follows: AC $= 66$ m, BC $= 40$ m, angle ACB $= 30°$. From these measurements, calculate the distance AB.

167

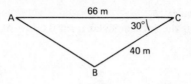

Fig. D6.11

See fig. D6.11:

$$c^2 = a^2 + b^2 - 2ab \cos C$$

$$\therefore \quad AB^2 = 40^2 + 66^2 - 2 \times 40 \times 66 \times \cos 30°$$

$$= 1600 + 4356 - 5280 \times 0.8660$$

$$= 5956 - 4572.48$$

$$= 1383.5$$

$$\therefore \quad AB = 37.2 \text{ m}$$

Example 2   Find the largest angle in a triangle with sides 3 m, 5 m, and 7 m.

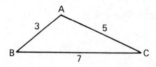

Fig. D6.12

The largest angle must be opposite the largest side, i.e. angle A opposite side BC in fig. D6.12.

$$\cos A = \frac{b^2 + c^2 - a^2}{2bc}$$

$$= \frac{25 + 9 - 49}{2 \times 3 \times 5} = \frac{-15}{30} = -\tfrac{1}{2}$$

$$\therefore \quad A = 120°$$

D6.5 Area of a triangle and of a segment

Area of a triangle by the semiperimeter formula
We have already seen (section D6.2) that, for any triangle ABC,

$$\text{area} = \tfrac{1}{2}bc \sin A = \tfrac{1}{2}ac \sin B = \tfrac{1}{2}ab \sin C$$

168

An alternative formula used to find the area when the three sides are known is in terms of the semiperimeter s:

$$\text{area} = \sqrt{s(s-a)(s-b)(s-c)}$$

where $s = \frac{1}{2}(a+b+c)$

Example For the triangle in fig. D6.12,

$$s = \frac{1}{2}(7+5+3) = 7.5$$

thus area $= \sqrt{7.5 \times 0.5 \times 2.5 \times 4.5}$

$$= 6.495 \text{ m}^2$$

Alternatively,

$$\text{area} = \frac{1}{2}bc \sin A = \frac{1}{2} \times 15 \sin 120^\circ$$

$$= 6.495 \text{ m}^2$$

Area of a segment

A segment of a circle is the portion between a chord and the circumference. The segment shown shaded in fig. D6.13 is bounded by the chord RS and the arc between R and S. This segment is the smaller of the two segments into which the circle is divided by chord RS. When such a chord divides a circle into two segments, the larger area is called the *major* segment and the smaller area is the *minor* segment. A diameter divides a circle into two equal segments and then each segment is a *semicircle*.

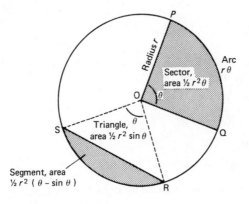

Fig. D6.13

To find the area of a segment (see fig. D6.13) we find the area of the sector (see section D5.6) and subtract from it the unwanted triangle. Using the formula $\frac{1}{2}ab \sin C$ for the area of the triangle, we have in this case

$$\text{area of triangle ORS} = \frac{1}{2}r^2 \sin \theta$$

169

thus   area of segment = sector − triangle

$$= \tfrac{1}{2}r^2\theta - \tfrac{1}{2}r^2 \sin\theta$$

$$= \tfrac{1}{2}r^2(\theta - \sin\theta)$$

For the use of the above formula, both the radius and the angle of the segment are required and must be derived from such measurements as can be made or may be supplied.

Example Find the area of brickwork necessary to fill the space between the soffit and the springing line of a segmental arch of span 18 m and rise 3 m.

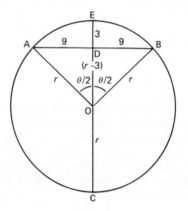

Fig. D6.14

The radius may be found by applying the theorem of Pythagoras to the triangle OBD in fig. D6.14:

$$r^2 = (r-3)^2 + 9^2$$

$$= r^2 - 6r + 9 + 81$$

$$\therefore \quad 6r = 90$$

$$r = 15\,\text{m}$$

The angle of the segment is found by trigonometry from the same triangle OBD:

$$\sin \tfrac{1}{2}\theta = \tfrac{9}{15} = \tfrac{3}{5} = 0.6000$$

$$\tfrac{1}{2}\theta = 36°52'$$

$$\theta = 73°44' = 1.2869 \text{ rad}$$

$$\sin \theta = 0.9600$$

170

Now area $A = \frac{1}{2}r^2(\theta - \sin\theta)$

$= \frac{1}{2}(15)^2(1.2869 - 0.9600)$

$= 36.78\,\text{m}^2$

Exercise D6

1 On a plan view of a high-speed locomotive, the windscreen appears to be 915 mm long. If the windscreen is inclined at an angle of 44° to the horizontal, what is its true length?

2 A lean-to greenhouse has a square horizontal base with sides 3.8 m in length. Estimate the area of glass needed to cover the roof, which is inclined at an angle of 32° to the base.

3 A straight line OP of length 100 mm is drawn on a graph from the origin to point P. If the angle between OP and the x-axis is 38°, find the projections of OP on to the x- and y-axes and hence determine the co-ordinates of point P.

4 Find the area of a triangle ABC in which BC = 320 mm, AC = 250 mm, and angle C is 37.5°.

5 A field forms a quadrilateral ABCD in which AB = 95 m, BC = 150 m, and CD = 80 m. If angle B is 90° and angle C is 112°, find the area of the field.

6 In triangle ABC, given that AB = 8 m, BC = 9 m, and angle A is 64°, find the size of angle C.

7 Given a triangle XYZ in which angle X is 58°, angle Y is 72°, and XY = 7.5 m, find the lengths of the other two sides.

8 From a triangle PQR in which QR = 7 km, RP = 17 km, and angle PQR is 125°, find angle QPR.

9 A body is supported by two chains connected to points 3 m apart on a horizontal beam. If the chain lengths are 1.8 m and 2.2 m, find the angle between the chains.

10 A ship leaves port and sails 50 km due N and then changes course to NE for a further 36 km. Find its actual distance from its starting point.

11 Adjacent sides of a parallelogram measure 15 mm and 19 mm, with an angle of 83.50° between them. Calculate the area of the parallelogram and the length of each diagonal.

12 A is the centre of a circle radius 3 m, B is the centre of a circle radius 4 m, and C is the centre of a circle radius 5 m. Find angles A, B, C, if each circle just touches the other two. Hence calculate the area in the centre of triangle ABC not covered by any of the three circles.

13 Figure D6.15 shows two positions of the jib of a crane (AC). Find by how much the cable (BC) has to be shortened to raise the jib from the first position to the second.

14 Use the cosine rule to prove that, for any parallelogram, the sum of the squares on the diagonals equals the sum of the squares on the four sides.

15 A space capsule is expected to return to earth and splash down in an area under observation from three points A, B, and C. Given the distances AB = 24 km, BC = 19 km, CA = 17 km, find the search area, triangle ABC.

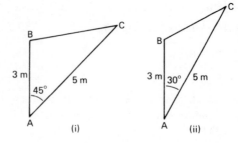

Fig. D6.15

16 Find the area of a segment of a circle radius 5.6 m for which the angle at the centre is 80°.

17 A chord of a circle radius 7 m passes exactly 1 m from the centre. Find the area of the two segments into which the circle is divided by the chord.

18 The cross-section of a passageway has two vertical sides each 3 m high, and a horizontal base 2 m long. The top is an arc of a circle. The centre of this circle is the centre of the rectangle formed by the base and sides. Find the area of the cross-section.

D7 Graphs of trigonometrical functions

D7.1 Plotting sine and cosine curves
We are already familiar with the shapes of the curves $y = \sin x$ and $y = \cos x$ for values of x between 0° and 360°, and with how they can be sketched by using projections of a moving radius on to the vertical and horizontal diameter of a circle, as shown in fig. D7.1. To sketch the curves for negative values of x, we can use the same method, except that the projections are made as the radius moves clockwise instead of anticlockwise.

Values for $\sin x$ and $\cos x$ for both $+$ and $-x$ can be obtained using a calculator. The table below shows values for both functions between $-360°$ and $+360°$

x	0	30	60	90	120	150	180	210	240	270	300	330	360
(degrees)	-360	-330	-300	-270	-240	-210	-180	-150	-120	-90	-60	-30	
$\sin x$	0	0.50	0.87	1	0.87	0.50	0	-0.50	0.87	-1	-0.87	-0.50	0
$\cos x$	1	0.87	0.50	0	-0.50	-0.87	-1	-0.87	-0.5	0	0.50	0.87	1

Figures D7.2 and D7.3 show the graphs $y = \sin x$ and $y = \cos x$ plotted over the range of angles in the table.

Comparing the two graphs on both sides of the line $x = 0°$, we see that the cosine curve for negative angles is a *mirror image* of the positive side. This

172

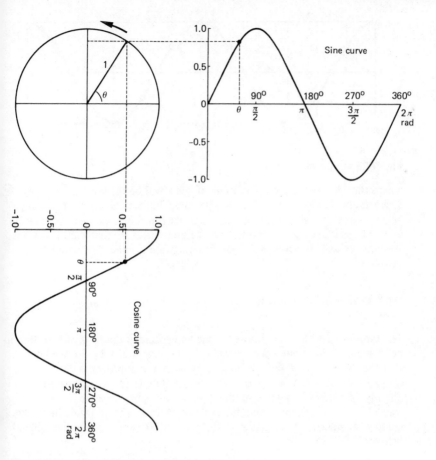

Fig. D7.1 Construction of sine and cosine curves

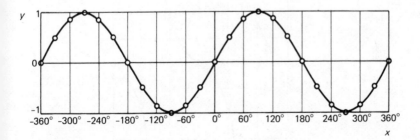

Fig. D7.2 Graph of $y = \sin x$

173

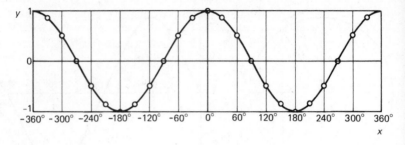

Fig. D7.3 Graph of $y = \cos x$

means that, for any angle x, the value of $\cos x$ will be the same as $\cos(-x)$. This relationship is not true for the sine curve, but we can say that, for any angle x, $\sin x = -\sin(-x)$. Functions like $\cos x$ in this respect are called *even*, while others like $\sin x$ are known as *odd* functions. Note that even functions are symmetrical about the axis; odd functions are symmetrical about the origin.

D7.2 Graphs of trigonometric functions

$y = \tan x$

The tangent curve is unlike those of sine and cosine in that it is not continuous; neither does it fluctuate between values of +1 and −1 as they do. We know that the values of $y = \tan x$ become very large as x approaches $90°$ ($\tan 89° = 57.29$). At $x = 90°$, $\tan x$ is infinitely large (written as ∞). Slightly beyond $90°$ its value suddenly becomes highly negative ($\tan 91° = -57.29$), increasing to zero at $x = 180°$. Figure D7.4 shows the curve $y = \tan x$ for the range $x = 0°$ to $x = 360°$, drawn using the table below:

x (degrees)	0	30	60	90	120	150	180	210	240	270	300	330	360
$\tan x$	0	0.58	1.73	α	−1.73	−0.58	0	0.58	73	∞	−1.73	−0.58	0

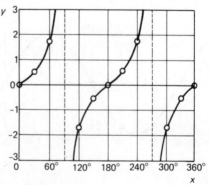

Fig. D7.4 Graph of $y = \tan x$

174

Comparing this curve with those of figs D7.2 and D7.3, when we imagine what it looks like on the negative side of $x = 0°$ it is clear that $y = \tan x$ is an odd function like $y = \sin x$; i.e. $\tan(-28°) = -\tan(28°) = -0.53$.

$y = \sin 2x$

To draw the sine curve in fig. D7.2 we simply plotted values of angle x against the corresponding $\sin x$ values obtained from a calculator or tables. If for values between $0°$ and $180°$ we now double each angle x to $2x$ and then look up the sine of $2x$ in each case, we produce the following table:

x (degrees)	0	15	30	45	60	75	90	105	120	135	150	165	180
$2x$ (degrees)	0	30	60	90	120	150	180	210	240	270	300	330	360
$\sin 2x$	0	0.5	0.87	1	0.87	0.5	0	−0.5	−0.87	−1	−0.87	−0.5	0

These results are shown plotted with remaining values to $x = 360°$ to give the graph $y = \sin 2x$ in fig. D7.5. This curve has the same sinusoidal shape as $y = \sin x$, but completes two cycles between $x = 0°$ and $x = 360°$ instead of one. The maximum and minimum values are again +1 and −1.

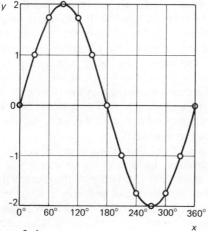

Fig. D7.5 Graph of $y = \sin 2x$

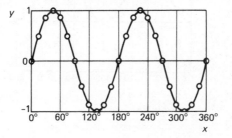

Fig. D7.6 Graph of $y = 2 \sin x$

175

$y = 2 \sin x$

By doubling values of $y = \sin x$ we used to plot fig. D7.2, the graph shown in fig. D7.6 for $y = 2 \sin x$ can be drawn. The maximum and minimum values are seen to be at $+2$ and -2, and only one cycle is completed in the first $360°$.

$y = \sin^2 x$

Squaring values of $y = \sin x$ for the range $x = 0°$ to $x = 360°$ gives the table below:

x (degrees)	0	30	60	90	120	150	180	210	240	270	300	330	360
$\sin x$	0	0.5	0.87	1	0.87	0.5	0	−0.5	−0.87	−1	−0.87	−0.5	0
$\sin^2 x$	0	0.25	0.75	1	0.75	0.25	0	0.25	0.75	1	0.75	0.25	0

Since all these values are positive, the curve (see fig. D7.7) remains above the x-axis over its whole length.

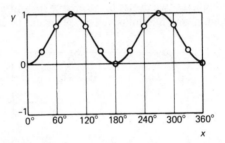

Fig. D7.7 Graph of $y = \sin^2 x$

D7.3 Curves of the form $a \sin nx$

In section D7.2 we saw that the curve $y = \sin 2x$ completed two cycles between $x = 0°$ and $x = 360°$. We would also find that a curve $y = \sin 3x$ completes three cycles in the first $360°$ if we drew it. In general, the curve $y = \sin nx$ will have n complete cycles in the first $360°$ or 2π radians. Looking at this another way, it means that the curve will repeat itself each time nx increases by $360°$. The fraction $360°/n$ or $2\pi/n$ is called the *period* of the waveform.

Also in the last section, we noted that the curve $y = 2 \sin x$ varies between $+2$ and -2. Similarly, a graph of $y = 5 \sin x$ would lie between $+5$ and -5. In the general case, the range of $y = a \sin x$ is $+a$ to $-a$. The quantity a is known as the *amplitude* of the waveform.

We can make use of these definitions of period and amplitude to sketch a curve for any equation of the form $y = a \sin x$ or $y = a \cos x$. Without a complete table of values, our knowledge of the sinusoidal shape of these curves, together with amplitude and period data, is sufficient to fill in the major details on a sketch. The curve $y = \frac{1}{2} \sin (3x/2)$ has amplitude $a = \frac{1}{2}$ and will range in value between $y = +\frac{1}{2}$ and $-\frac{1}{2}$. Also, since $n = \frac{3}{2}$, the period of this curve is $360°/\frac{3}{2} = 240°$ (i.e. a complete cycle every $240°$). Figure

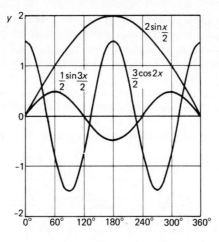

Fig. D7.8 Graph of the curves $y = \frac{1}{2}\sin(3x/2)$, $y = \frac{3}{2}\cos 2x$, and $y = 2\sin(x/2)$

D7.8 shows this curve, together with those of $y = \frac{3}{2}\cos 2x$ (which has amplitude $\frac{3}{2}$, period $180°$) and $y = 2\sin(x/2)$ (which has amplitude 2, period $720°$).

D7.4 Combination of waveforms

There are many instances in engineering when it is useful to study what happens when waveforms combine; for example, when dealing with alternating currents, mechanical vibrations, or the interaction of sound waves. Figure D7.9 shows the curves $y = \sin x$ and $y = 2\sin x$ drawn to the same scale (broken lines). We can add them together graphically by simply taking the y-values of both curves at a number of points along the x-axis and adding them together.

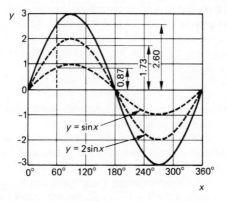

Fig. D7.9 Graphical addition of $y = \sin x$ and $y = 2\sin x$

177

Next, we mark a new point on the graph with y-value or *ordinate* equal to the sum of those values from the two curves. Consider the point A in fig. D7.9 at $x = 60°$. At this point, $\sin x = 0.87$ and $2 \sin x = 1.73$. The new point is therefore at $y = 0.87 + 1.73 = 2.60$. Repeating this procedure for a range of x-values gives the curve $y = 3 \sin x$.

Using the same method we can also add together two cosines. Furthermore, subtraction may be carried out graphically with both sine and cosine curves.

Example Show graphically the subtraction $\cos x - 3 \cos x$ from $x = 0°$ to $x = 360°$.

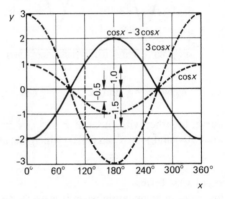

Fig. D7.10 Graphical subtraction: $\cos x - 3 \cos x$

First we draw the two curves to be subtracted, to the same scale, over the range of x-values specified (shown as broken lines in fig. D7.10).

We must take care to subtract the ordinates of the curve $y = 3 \cos x$ from those of $y = \cos x$ correctly. This is especially so when they are negative, as happens for values between $x = 90°$ and $x = 270°$. At the point in fig. D7.10 when $x = 120°$, $3 \cos x = -1.5$ and $\cos x = -0.5$. Subtracting these gives $(-0.5) - (-1.5)$ or $(-0.5) + (1.5) = +1$. The point is plotted at $y = 1$ on the resultant curve (solid line), which is $y = -2 \cos x$.

To generalise, we have seen that, when we combine two sine curves or two cosine curves with the same period, the resultant is a curve of the same type with the same period. The resultant amplitude will be the *sum* of the amplitudes of the component curves when they are added, and their *difference* if we subtract the curves.

We now look at what happens when a sine wave is combined with a cosine wave with the same period. Figure D7.11 shows the graphical addition of the curves $y = 2 \sin x$ and $y = \cos x$ (both drawn with broken lines). The resultant curve (drawn as in previous cases by adding ordinates of the original

178

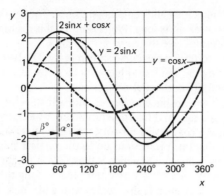

Fig. D7.11 Graphical addition: $2 \sin x + \cos x$

curves), although sinusoidal in shape, does not have a zero value at $x = 0°$.
We see from fig. D7.11 that it reaches a maximum value $\alpha°$ before the point
$x = 90°$. When a sine curve is moved to the *left* in this way, the lateral displacement $\alpha°$ is a *phase angle* or *phase lead*. If the waveform is displaced to
the *right*, the phase angle is negative. This means that we could also describe
the resultant curve in fig. D7.11 as a cosine curve with phase angle $-\beta°$. We
say the curve has a $\beta°$ *phase lag*.

D7.5 Using a resultant phasor to draw combined waveforms
Another method of drawing the resultant curve formed by the combination
of two waveforms involves projections of a rotating radius or *phasor*. We must
first draw the *resultant phasor* by combining those of the waveforms. Phasors
are added in the same way as vectors (students unfamiliar with the addition
of vectors should read section D9).

Consider the addition $\sin x + 2 \sin x$ already dealt with in section D7.4 (see
fig. D7.12). We add phasor OP, which is one unit long and represents $\sin x$, to

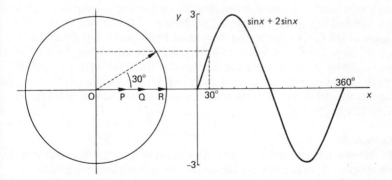

Fig. D7.12 Projections of a resultant phasor

OQ which is two units long and represents 2 sin x. The resultant phasor OR (see fig. D7.12) is three units long. Projections of this resultant on to the vertical axis as it rotates give the curve $y = 3 \sin x$.

You should try to obtain the curve shown in fig. D7.10 for the subtraction $\cos x - 3 \cos x$ using the resultant-phasor method, by relating the angle of a rotating phasor with its horizontal projections.

With mixed sine and cosine curves, the resultant waveform will have a phase lead or lag (see fig. D7.11). To determine the waveform resulting from the addition $2 \sin x + \cos x$ using phasors, we again relate the angle of a resultant rotating phasor and its projections on to the vertical axis. In fig. D7.13 we represent $2 \sin x$ by a phasor OP which is drawn with length two units along the horizontal axis. We draw $\cos x$ as a unit phasor along the vertical axis. (Note we can treat all cosines as sine curves with a 90° phase lead, i.e. $\cos x = \sin (x + 90)°$, by drawing phasors along the vertical axis rather than the horizontal axis.) Phasor OR is the resultant obtained from the horizontal and vertical components OP and OQ.

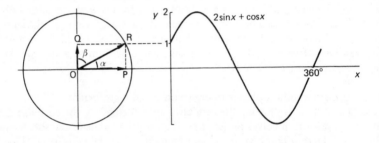

Fig. D7.13 Projections of a resultant phasor to draw $2 \sin x + \cos x$

Note that the curve drawn using projections of rotating phasor OR on to the vertical axis is the same as we obtained by plotting the curves $2 \sin x$ and $\cos x$ and adding them together. The amplitude of the curve is equal to the length of OR (2.24 units by the theorem of Pythagoras), and the phase lead α is the angle which OR makes with the horizontal axis before the start of anticlockwise rotation. The curve is defined as OR $\sin (x + \alpha)$. The angle β which OR makes with the vertical axis in fig. D7.13 corresponds to the phase lag when we treat the resultant waveform as a cosine curve. In this case we write OR $\cos (x - \beta)$.

Exercise D7
1   Draw a table of values for $y = \sin x$ between $x = 0$ and π radians with points every $\pi/10$ rad and use these values to draw an accurate graph.
2   Plot the curve $y = \tan x$ between $x = 0°$ and $x = 90°$ using a vertical axis with maximum value $y = 10.00$. Show points every 10° along the x-axis.

3 Plot an accurate graph of $y = \cos 2x$ between $x = -90°$ and $x = +90°$, using values of x at $15°$ intervals.

4 Write down the period and amplitude of the waveforms (a) $y = \cos 3.6x$, (b) $y = \sqrt{3} \sin 2\pi x$, (c) $y = \frac{1}{2} \sin 6x$.

5 Plot the curves $y = \cos^2 x$ and $y = 2 \cos x$ on the same graph, using points every $30°$ between $x = 0°$ and $x = 360°$. How many times do the curves cross?

6 Sketch the curve $y = \frac{1}{2} \sin (2x/3)$ for one cycle.

7 Draw the curves $y = 2 \cos x$ and $y = 0.8 \sin x$ on the same graph over one complete period. Add them together and measure the amplitude of the resultant and also the point where it first crosses the x-axis.

8 Show graphically by the subtraction of two waveforms the curve $y = 3 \sin x - 2 \cos x$. Measure the amplitude and phase angle of this sine wave.

9 Use projections from a resultant phasor on to the vertical axis to draw the combined waveform $\sin x + 2.5 \cos x$. Measure the amplitude and phase lead of the curve.

D8 Trigonometrical relationships and equations

D8.1 Deriving simple relationships

The six trigonometrical ratios have been defined in section D5.1 in terms of the sides of a right-angled triangle such as ABC in fig. D8.1.

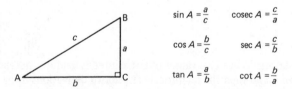

$$\sin A = \frac{a}{c} \qquad \operatorname{cosec} A = \frac{c}{a}$$

$$\cos A = \frac{b}{c} \qquad \sec A = \frac{c}{b}$$

$$\tan A = \frac{a}{b} \qquad \cot A = \frac{b}{a}$$

Fig. D8.1 The six trigonometrical ratios

It is clear that these ratios are not independent but can be related; for example by using the reciprocals:

$$\operatorname{cosec} A = \frac{1}{\sin A} \qquad \sec A = \frac{1}{\cos A} \qquad \cot A = \frac{1}{\tan A}$$

We can also take the ratio $\tan A = a/b$ and divide both numerator and denominator of this fraction by c, giving

181

$$\tan A = \frac{a/c}{b/c} = \frac{\sin A}{\cos A}$$

It follows from this that

$$\cot A = \frac{1}{\tan A} = \frac{\cos A}{\sin A}$$

Example 1   Given that $\sin x = 0.6000$ and $\cos x = 0.8000$, find the values of $\tan x$ and $\cot x$.

First method

$$\tan x = \frac{\sin x}{\cos x} = \frac{0.6}{0.8} = 0.7500$$

$$\cot x = \frac{1}{\tan x} = \frac{1}{0.7500} = 1.3333$$

Second method   Sketch a small diagram (see fig. D8.2) to show a right-angled triangle and insert known values, from which

$$\tan x = \frac{0.6}{0.8} = 0.7500$$

$$\cot x = \frac{0.8}{0.6} = 1.3333$$

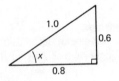

Fig. D8.2

Many calculators do not give values of $\cot x$ directly and it is necessary to find first $\tan x$ and then the reciprocal. The above derivations using the ratios of sides of a right-angled triangle are obviously true for any angle between $0°$ and $90°$. We can verify numerically that they hold true for all angles. If we let $x = 155°$, for example, $\sin 155° = 0.4226$ and $\cos 155° = -0.9063$. Dividing these gives

$$\tan x = \frac{0.4226}{-0.9063} = -0.4663$$

which when we check is $\tan 155°$.

Relationships such as these which are true for any angle are called *identitie* and can often be used to simplify trigonometrical expressions in practical problems.

Example 2 Prove that cosec θ tan θ = sec θ.

We can substitute for the trigonometrical ratios on the l.h.s. and then cancel as follows:

$$\frac{1}{\cancel{\sin\theta}} \cdot \frac{\cancel{\sin\theta}}{\cos\theta} = \frac{1}{\cos\theta} = \sec\theta$$

D8.2 $\sin^2 A + \cos^2 A = 1$ and other identities

Applying the theorem of Pythagoras to fig. D8.1 in section D8.1,

$$a^2 + b^2 = c^2$$

Dividing both sides of this equation by c^2 gives

$$\left(\frac{a}{b}\right)^2 + \left(\frac{b}{c}\right)^2 = 1$$

We know that in triangle ABC, $\sin A = a/c$ and $\cos A = b/c$, so $(a/c)^2 = (\sin A)^2$ or, as it is usually written, $\sin^2 A$. Similarly, $(b/c)^2 = \cos^2 A$ and our equation becomes

$$\sin^2 A + \cos^2 A = 1$$

Taking as an example $A = 44°$, we see that $\sin A = \sin 44° = 0.6947$ and so $\sin^2 A = (0.6947)^2 = 0.4826$. Similarly, $\cos^2 A = 0.5174$. Adding these values confirms the above relationship:

$$\sin^2 A + \cos^2 A = 0.4826 + 0.5174 = 1$$

If we divide both sides of the relationship by $\cos^2 A$ we see that

$$\frac{\sin^2 A}{\cos^2 A} + \frac{\cos^2 A}{\cos^2 A} = \frac{1}{\cos^2 A}$$

or $\qquad \tan^2 A + 1 = \sec^2 A$

Dividing by $\sin^2 A$ instead of $\cos^2 A$ gives

$$1 + \cot^2 A = \csc^2 A$$

As with the relationships derived in section D8.1, we can show that these three equations are true for any angle and are therefore identities. If the first of them is remembered, it is easy to derive the other two when required.

Example 1 Show that $\cos^2\theta - \sin^2\theta = 2\cos^2\theta - 1$.

Using the identity $\sin^2 A + \cos^2 A = 1$, we can substitute $\sin^2\theta = 1 - \cos^2\theta$ in the l.h.s. of the equation to give

$$\cos^2\theta - (1 - \cos^2\theta) = 2\cos^2\theta - 1$$

Alternatively,

$$\cos^2\theta - \sin^2\theta = 2\cos^2\theta - 1$$

Add $\quad \dfrac{\sin^2\theta + \cos^2\theta = 1}{\cos^2\theta + \cos^2\theta = 2\cos^2\theta}$

Example 2 Show that $(1 - \sec\theta)(1 + \sec\theta) + \tan^2\theta = 0$.

Multiplying the brackets gives on the l.h.s.

$$1 - \sec^2\theta + \tan^2\theta$$

From the identity $\tan^2 A + 1 = \sec^2 A$, we can write

$$1 - \sec^2\theta = -\tan^2\theta$$

Substituting in the equation gives

$$-\tan^2\theta + \tan^2\theta = 0$$

Alternatively,

$$1 - \sec^2\theta + \tan^2\theta = 0$$

Add $\qquad \dfrac{\sec^2\theta \quad = 1 + \tan^2\theta}{1 + \tan^2\theta = 1 + \tan^2\theta}$

D8.3 Solving trigonometrical equations

$a \sin\theta = b$

A trigonometrical equation, unlike most algebraic equations, will usually have an infinite number of solutions; for example, the solutions of the equation $\sin\theta = 0.5$ include $\theta = 30°, \theta = 150°, \theta = 390°,$ and $\theta = 510°$, as well as negative angles $\theta = -210°, \theta = -330°$. When we sketch the curve $y = \sin\theta$ (see fig. D8.3), we see that the solutions are all the values of θ at which a line drawn horizontally through $y = 0.5$ crosses the sine curve.

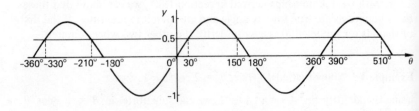

Fig. D8.3 Some solutions of the equation $\sin\theta = 0.5$

In order to solve these equations, a thorough knowledge of the nature of the basic trigonometrical ratios is necessary. Given an equation of the type $\sin \theta = a$, $\cos \theta = a$, or $\tan \theta = a$, we can find an initial value of θ between $-180°$ and $+180°$ using a calculator or a book of tables. From this value, further solutions can be written down. A sketch of the trigonometrical function is most helpful in locating other solutions from the symmetry and regularity of the curve.

Example 1   Find the solutions of the equation $\cos \theta = 0.73$ which lie between $\theta = 0°$ and $\theta = 360°$.

From a calculator, arccos $0.73 = 43.11°$.

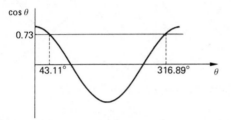

Fig. D8.4 Solutions of the equation $\cos \theta = 0.73$

A sketch of the curve $y = \cos \theta$ between $0°$ and $360°$ (see fig. D8.4) shows that there will be only one other solution to the equation in the specified range. From the symmetry of the curve, this second value will be

$$\theta = 360° - 43.11°$$
$$= 316.89°$$

i.e. the required solutions are $\theta = 43.11°$ and $316.89°$.

It is sometimes useful to work in radians instead of degrees. Most calculators will do this automatically and you should find out how to operate the calculator you use in this way. When direct operation in radians is not possible, we can convert from degrees to radians simply by dividing an answer in degrees by 57.296 (since π radians $= 180°$, 1 radian $= 57.296°$).

Example 2   Find, in radians, the values of θ between 0 and 2π radians which satisfy the equation $2 \sin \theta = -0.56$.

Dividing by 2 gives

$$\sin \theta = -0.28$$

Using a calculator set to work in radians to find the first solution of this equation,

$$\theta = \arcsin(-0.28) = -0.284 \text{ rad}$$

185

Working in degrees to solve the same problem,

$$\theta = \arcsin(-0.28) = -16.26°$$

$$= \frac{-16.26}{57.3}\text{ rad}$$

$$= -0.284\text{ rad}$$

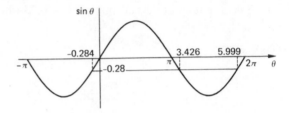

Fig. D8.5 Solutions of the equation $\sin\theta = -0.28$

A sketch (see fig. D8.5) shows which solution this is. Although not in the required range, we can use this value and the curve to find two solutions which are between 0 and 2π radians. These are

$$\theta = \pi + 0.284$$

$$= 3.142 + 0.284$$

$$= 3.426\text{ rad}$$

and $\quad\theta = 2\pi - 0.284$

$$= 6.283 - 0.284$$

$$= 5.999\text{ rad}$$

i.e. the solutions are $\theta = 3.426$ rad and 5.999 rad.

$a\sin^2\theta = b$
When an equation involves squared trigonometrical terms, we must always include plus and minus signs when we take a square root; possible solutions to the equation will otherwise be missed.

Example 3   Find all solutions between $\theta = 0°$ and $\theta = 360°$ (often written as $0° \leqslant \theta \leqslant 360°$) of the equation $\sin^2\theta = 0.42$.

The solutions will be points at which a line drawn horizontally at $y = 0.42$ crosses the curve $y = \sin^2\theta$ within the range of angles specified. We can, as in the previous example, simplify calculation by reducing the equation to the form $\sin\theta = $ a constant.

186

Taking the square root of both sides of the original equation,

$$\sin \theta = \pm 0.648$$

Using a calculator,

$$\theta = \arcsin 0.648 = 40.39°$$

and $\theta = \arcsin(-0.648) = -40.39°$

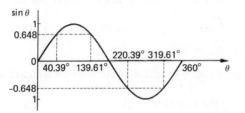

Fig. D8.6 Solutions of the equation $\sin \theta = \pm 0.648$

A sketch of the curve $y = \sin \theta$ (see fig. D8.6) shows that there are four solutions between $\theta = 0°$ and $\theta = 360°$. They are values of θ at which lines drawn horizontally at $+0.648$ and -0.648 cross the curve. Two are between $0°$ and $180°$ and, of these, one has already been given by the calculator as $\theta = 40.39°$. The second of these solutions is

$$\theta = 180° - 40.39°$$

$$= 139.61°$$

The two solutions between $180°$ and $360°$ are

$$\theta = 180° + 40.39° = 220.39°$$

and $\theta = 360° - 40.39° = 319.61°$

The four solutions of the equation $\sin^2 \theta = 0.42$ in the given range are therefore $\theta = 40.39°$, $139.61°$, $220.39°$, and $319.61°$.

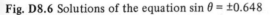

$a \sin^2 \theta + b \sin \theta + c = 0$

Quadratic equations involving trigonometrical functions can be solved in two stages. First, we use the usual methods (factorising or formula) to find roots. These will be in the form $\sin \theta = x$ or $\sin \theta = y$. The second stage is to find all solutions within a given range of angles from *both* roots.

Consider the equation

$$6 \sin^2 \theta - 7 \sin \theta + 2 = 0$$

187

This is a quadratic equation and will factorise to give

$$(2 \sin \theta - 1)(3 \sin \theta - 2) = 0$$

The roots are therefore

$$\sin \theta = \tfrac{1}{2} \text{ or } \sin \theta = \tfrac{2}{3}$$

Both of these roots are themselves trigonometrical equations of the type solved earlier in this section. They both have an infinite number of solutions. To find all solutions between $\theta = 0°$ and $\theta = 360°$, we take each root in turn.

From $\sin \theta = \tfrac{1}{2}$, we use a calculator (if necessary) to give

$$\theta = \arcsin 0.5 = 30°$$

Within the range of angles specified, there will be a second solution, at $\theta = 150°$.

Similarly, from the root $\sin \theta = \tfrac{2}{3}$,

$$\theta = \arcsin (2/3) = 41.81°$$

A second relevant value is $\theta = 180° - 41.81° = 138.19°$. Solutions of the original quadratic equation are therefore $\theta = 30°, 41.81°, 138.19°,$ and $150°$

Note that it is usual to put solutions in order of size.

Example 4   Find all solutions in the range $0° \leqslant x \leqslant 360°$ of the equation $\cos^2 x - 1.19 \cos x + 0.27 = 0$.

Using the formula for the solution of $ax^2 + bx + c = 0$:

$$x = \frac{-b \pm \sqrt{(b^2 - 4ac)}}{2a}$$

we have $a = 1, b = -1.19,$ and $c = 0.27$

$$\therefore \quad \cos x = \frac{1.19 \pm \sqrt{[(-1.19)^2 - 4 \times 1 \times 0.27]}}{2}$$

$$= \frac{1.19 \pm 0.580}{2}$$

The roots are $\cos x = 0.885$ and $\cos x = 0.305$. Taking the first of these, we find one solution using a calculator:

$$x = \arccos 0.885 = 27.75°$$

From this, another solution in the range is

$$x = 360° - 27.75° = 332.25°$$

Taking the second root, $\cos x = 0.305$,

$$x = \arccos (0.305) = 72.24°$$

and the second value in the given range is

$$x = 360° - 72.24° = 287.76°$$

There are, therefore, four solutions to the quadratic equation between $x = 0°$ and $x = 360°$. They are $x = 27.75°, 72.24°, 287.76°$, and $332.25°$.

D8.4 The use of identities to simplify trigonometrical equations
In some cases it may be necessary to use the identities derived in sections D8.1 and D8.2 to simplify trigonometrical equations before we solve them. For example, to find values of θ (between $0°$ and $360°$) which satisfy the equation

$$4 \sin \theta = 5 \cos \theta$$

we can divide both sides by $4 \cos \theta$ to give

$$\frac{\sin \theta}{\cos \theta} = \frac{5}{4}$$

and, using the identity $\tan \theta = \sin \theta / \cos \theta$, the equation becomes

$$\tan \theta = \tfrac{5}{4}$$

By finding solutions of this equation we can solve the original one.
    Using a calculator,

$$\theta = \arctan \left(\tfrac{5}{4}\right) = 51.34°$$

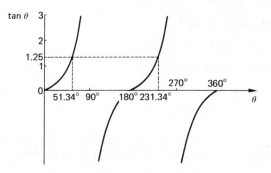

Fig. D8.7 Solutions of the equation $\tan \theta = \tfrac{5}{4}$

A sketch of the curve $y = \tan \theta$ (see fig. D8.7) shows that there is a second value of θ within the given range of angles $0°$ to $360°$. The second solution is

$$\theta = 180° + 51.34° = 231.34°$$

Checking these solutions in the original equation,

$$4 \sin 51.34° = 3.12 = 5 \cos 51.34°$$

and $\quad 4 \sin 231.34° = -3.12 = 5 \cos 231.34°$

These two solutions are, in fact, angles at which the curves $y = 4 \sin \theta$ and $y = 5 \cos \theta$ cross.

Example Find values of θ in the range $0° \leqslant \theta \leqslant 360°$ which satisfy the equation

$$\frac{2 \sin^2\theta}{(1 + \cos \theta)^2} = 1$$

Multiplying both sides of the equation by $(1 + \cos \theta)^2$,

$$2 \sin^2\theta = (1 + \cos \theta)^2$$
$$= 1 + \cos^2\theta + 2 \cos \theta$$

Using the identity $\sin^2 A + \cos^2 A = 1$, the equation can be written

$$2(1 - \cos^2\theta) = 1 + \cos^2\theta + 2 \cos \theta$$

or $3 \cos^2\theta + 2 \cos \theta - 1 = 0$

The quadratic expression on the l.h.s. factorises:

$$(3 \cos \theta - 1)(\cos \theta + 1) = 0$$

and the roots are therefore

$$\cos \theta = \tfrac{1}{3} \quad \text{and} \quad \cos \theta = -1$$

From the first root, using a calculator,

$$\theta = \arccos \left(\tfrac{1}{3} \right) = 70.53°$$

A second solution in the range $0° \leqslant \theta \leqslant 360°$ is

$$\theta = 360° - 70.53° = 289.47°$$

From the second root,

$$\theta = \arccos (-1) = 180°$$

This is the only value in the given range. The solutions to the original trigonometrical equation are therefore $\theta = 70.53°, 180°,$ and $289.47°$.

Exercise D8
1   By making both sides the same, show that the following equations are identities:

a) $\operatorname{cosec} \theta = \dfrac{\cot \theta}{\cos \theta}$

b) $\sin^2 A \sec A = \sin A \tan A$

c) $\dfrac{\operatorname{cosec} x}{\tan x} = \dfrac{\cos x}{\sin^2 x}$

2 Given that $\sin x = 0.993$ and $\cos x = 0.118$, find $\tan x$, $\operatorname{cosec} x$, and $\operatorname{cosec}^2 x$, using multiplication and division only.

3 If $\sin A \cos A = 0.5$, show that $(\sin A - \cos A)^2 = 0$.

4 Simplify the following expressions:

a) $\dfrac{1 + \tan^2\theta}{1 + \cot^2\theta}$

b) $\sqrt{\dfrac{\sin^2 x(1 - \operatorname{cosec}^2 x)}{(\cos^2 x - 1)}}$

5 Show that $(\sin\theta - \operatorname{cosec}\theta)^2 = \cot^2\theta - \cos^2\theta$.

6 Find the values of x between $-360°$ and $360°$ which satisfy the equations (a) $\sin x = 0.40$, (b) $\cos x = -0.86$, (c) $\tan x = 2.48$.

7 Solve the equation $3.5 = 4.7 \sin x$, giving all solutions between $0°$ and $180°$.

8 Find all values between $\theta = -180°$ and $\theta = 180°$ which satisfy the equation $12.6 \sin\theta = 3.2 \cos\theta$.

9 Find, in radians, all solutions of the equation $1/\sin\theta = -2.47$ between 0 and 2π rad.

10 Find all solutions in the range $-180° \leqslant A \leqslant 180°$ of the equation $4 \sin^2 A = 1.7$.

11 What are the values of x in the range $x = 0°$ to $x = 360°$ which satisfy the equation $5 \sin^2 x = 8 \cos^2 x$?

12 How many solutions of the equation $15 \sin^2\theta - 11 \sin\theta + 2 = 0$ are there between $\theta = -360°$ and $\theta = 360°$?

13 Find all solutions between $x = 0°$ and $x = 360°$ of the equation $6 \cos^2 x - 5 \cos x + 1 = 0$.

14 What values of x between $x = 0°$ and $x = 180°$ satisfy the equation $(\tan x - 5)^2 = 49$?

15 Find the four values of θ in the range $-180° \leqslant \theta \leqslant 180°$ which satisfy the equation $\sin^2\theta - 0.46 \sin\theta = 0.11$.

D9 Vectors

D9.1 What is a vector?

A vector quantity is one which has both *magnitude* and *direction*; for example, force, velocity, acceleration, and momentum are all vector quantities. *Scalar* quantities possess only magnitude. Temperature, frequency, mass, length, and time are all scalars.

A vector can be shown as a line drawn between two points. The length of the line gives the magnitude of the vector, and its direction is measured with reference to a set of axes. Two axes are used to describe vectors which are *coplanar* (i.e. in the same plane). Three axes are required to define vectors in three-dimensional space, but we shall limit our consideration to coplanar vectors only.

Figure D9.1(a) shows a line drawn from the origin O to a point A, and this line represents a vector drawn to scale. The arrowhead on the line indicates that the line runs from O to A and not from A to O. We use the symbol $\overline{OA}$ to represent the vector (in textbooks you will sometimes also see vectors indicated by bold sloping type, e.g. *OA*). The order of letters tells us the direction of the vector.

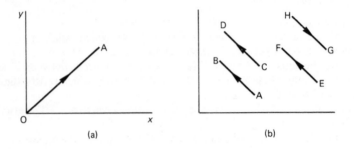

Fig. D9.1

Two vectors which have the same magnitude and direction are equal. Thus in fig. D9.1(b), $\overline{AB} = \overline{CD} = \overline{EF}$ and, since $\overline{HG}$ has the same magnitude but opposite direction, $\overline{HG} = -\overline{AB}$.

Vectors can be added and subtracted and − since they can be used to represent forces, velocities, and other engineering quantities − they provide a powerful method of analysing many mechanical and electrical systems.

D9.2 Vector addition
If a vector $\overline{OP}$ is added to $\overline{PQ}$, the *resultant* vector is $\overline{OQ}$. The addition of these vectors is shown in fig. D9.2, and we write

$$\overline{OP} + \overline{PQ} = \overline{OQ}$$

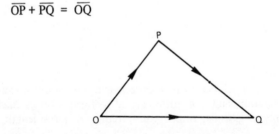

Fig. D9.2 The addition of vectors $\overline{OP}$ and $\overline{PQ}$

This does not mean that the lengths of lines OP and PQ add up to the length of OQ (unless, of course, $\overline{OP}$ and $\overline{PQ}$ are in a straight line).

192

If the sum of two vectors is zero, it means that they have the same magnitude and opposite directions.

When more than two vectors are added together, a polygon is drawn by joining vectors end to end in any order. The vector represented by a line joining the two free ends is the resultant vector. When the two ends of the chain of vectors meet, the resultant is zero and this condition represents a state of equilibrium in many engineering applications. Figure D9.3 shows the vector addition $\overline{AB} + \overline{BC} + \overline{CD} + \overline{DE} = \overline{AE}$.

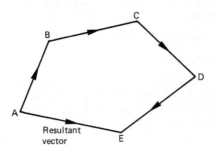

Fig. D9.3 Addition of four vectors

D9.3 Resolution of vectors

To calculate the magnitude and direction of a resultant vector, we start by *resolving* the vectors to be added into *component* parts at right angles (usually parallel to x- and y-axes). Figure D9.4 shows the horizontal and vertical component parts of a vector $\overline{OP}$. These are obtained by simply drawing lines perpendicular to the x- and y-axes from the end of the vector.

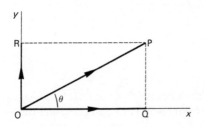

Fig. D9.4 Horizontal and vertical components of a vector

The horizontal component of $\overline{OP}$ is $\overline{OQ}$ and the vertical component is OR. Note that these components are themselves vectors and obey the rule of addition:

$$\overline{OQ} + \overline{OR} = \overline{OP}$$

193

They correspond in length to the Cartesian co-ordinates of the point P and may be found using trigonometry as follows. In fig. D9.4,

$$\cos \theta = \frac{OQ}{OP}$$

$\therefore \qquad OQ = OP \cos \theta$

also $\quad OR = QP = OP \sin \theta$

In general, we may resolve a vector $\overline{OP}$ at an angle θ to the horizontal into two components, where

$$\text{horizontal component} = OP \cos \theta$$

$$\text{vertical component} = OP \sin \theta$$

Note that we take horizontal components to be negative in the second and third quadrants and vertical components to be negative in the third and fourth quadrants; to avoid mistakes in your calculations you should always use the angle which the vector makes with the *positive* horizontal axis. Your calculator will give you the values of $\cos \theta$ and $\sin \theta$ with the correct sign.

Example 1   What are the x- and y-components of the vector $\overline{OA}$ joining the origin to the point A at $(3, -2)$?

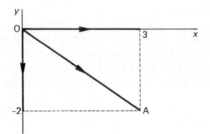

Fig. D9.5

Figure D9.5 shows the resolution of vector $\overline{OA}$ into components in the x- and y-directions. The x-component is $+3$ units long and the y-component is -2 units.

Example 2   Resolve into horizontal and vertical components a force of 10 N acting at an angle of $30°$ upwards from the ground.

The vector $\overline{OF}$ representing the force in magnitude and direction is shown in fig. D9.6. The horizontal and vertical components can be found graphically by drawing vector $\overline{OF}$ to scale using a ruler and protractor (with a scale, say, $1 \, N \equiv 1 \, cm$) and then completing rectangle OXFY to measure OX and OY.

194

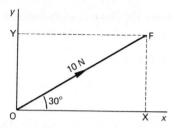

Fig. D9.6

Alternatively, we can calculate the components OX and OY as follows.

$$OX = OF \cos 30°$$
$$= 10 \times 0.866$$
$$= 8.66$$

∴ the horizontal component is 8.66 N.

$$OY = 10 \sin 30°$$
$$= 5.0$$

∴ the vertical component is 5 N.

D9.4 Calculating the resultant from component parts
If we reverse the process developed in section D9.3, we can determine the magnitude and direction of a resultant vector given its components at right angles.

Example Calculate the magnitude and direction of velocity vector $\overline{OP}$ which has a component along the x-axis $\overline{OX}$ = 3 m/s and along the y-axis $\overline{OY}$ = 2 m/s.

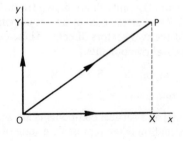

Fig. D9.7

We draw the vector $\overline{OP}$ as shown in fig. D9.7, with x- and y-axis components $\overline{OX}$ and $\overline{OY}$.

195

From triangle OXP, using the theorem of Pythagoras,

$$OP^2 = OX^2 + XP^2$$

Since $XP = OY = 2$ and $OX = 3$,

$$OP^2 = 9 + 4$$

$$\therefore \quad OP = 3.61 \text{ m/s}$$

To find the direction of $\overline{OP}$ we use the same triangle OXP:

$$\tan O = \tfrac{2}{3}$$

$$\therefore \quad \angle O = \arctan 0.67$$

$$= 33.69°$$

i.e. the resultant velocity vector has magnitude 3.61 m/s and is directed at an angle of 33.69° measured anticlockwise from the x-axis.

D9.5 Resolving a system of vectors

We can obtain total horizontal and vertical components of a system of vectors if we follow the procedure in section D9.3 to resolve each of the vectors horizontally and vertically and then add all components in the same direction.

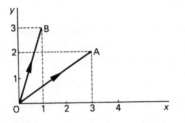

Fig. D9.8

In fig. D9.8, two vectors $\overline{OA}$ and $\overline{OB}$ are drawn from the origin to point A at $(3, 2)$ and B at $(1, 3)$. The Cartesian co-ordinates of points A and B give the x- and y-components of the two vectors directly. These can be added to give total horizontal and vertical components:

$$\text{total horizontal component} = 3 + 1 = 4$$

$$\text{total vertical component} = 3 + 2 = 5$$

When the total horizontal and vertical components are zero there will be no resultant vector. This situation often represents a state of *equilibrium* in engineering applications.

Example Determine the total horizontal and vertical components for the three forces in fig. D9.9 represented by vectors $\overline{OP}$, $\overline{OQ}$, and $\overline{OR}$.

196

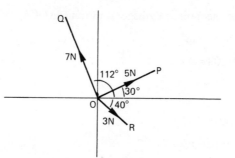

Fig. D9.9

The components of the vector $\overline{OP}$ are $5 \cos 30°$ horizontally and $5 \sin 30°$ vertically. We can find the total of all horizontal and vertical components:

$$\text{total horizontal component} = 5 \cos 30° + 3 \cos (-40°) + 7 \cos 112°$$
$$= 4.33 + 2.30 - 2.62$$
$$= 4.01 \text{ N}$$
$$\text{total vertical component} = 5 \sin 30° + 3 \sin (-40°) + 7 \sin 112°$$
$$= 2.50 - 1.93 + 6.49$$
$$= 7.06 \text{ N}$$

i.e. the system of vectors has a total horizontal component of 4.01 N and a total vertical component of 7.06 N.

D9.6 From total components to a resultant vector

Using the procedure in section D9.5, we have resolved a set of vectors into total horizontal and vertical components. If we now find the vector which has these components it will be the resultant of the original set of vectors. By drawing a rectangle OGSH (see fig. D9.10(a)), we can calculate the magnitude and direction of $\overline{OS}$, which is the resultant vector.

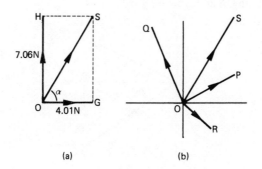

(a)   (b)

Fig. D9.10

197

Applying the theorem of Pythagoras to triangle OGS,

$$OS^2 = OG^2 + GS^2$$
$$= 16.08 + 49.84$$
$$\therefore \quad OS = 8.12\,N$$

Also $\tan \alpha = \dfrac{7.06}{4.01}$

$$\therefore \qquad \alpha = \arctan 1.76$$
$$= 60.40°$$

Figure D9.10(b) shows the resultant vector $\overline{OS}$ (magnitude 8.12 N inclined at 60.40° to the horizontal) in relation to the three original vectors $\overline{OP}$, $\overline{OQ}$, and $\overline{OR}$.

Example   Figure D9.11 shows two tugs attached by cables to an oil rig. If tug A exerts a force of 250 kN and B exerts 350 kN in the directions indicated, calculate the magnitude and direction of the resultant force on the rig.

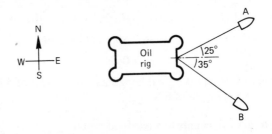

Fig. D9.11

We represent forces of tugs A and B by vectors $\overline{OA}$ and $\overline{OB}$ and resolve them into horizontal and vertical components (see fig. D9.12). The horizontal components of vectors $\overline{OA}$ and $\overline{OB}$ are $\overline{OC}$ and $\overline{OD}$ respectively, and the vertical components are $\overline{CA}$ and $\overline{DB}$.

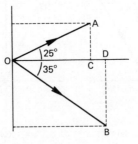

Fig. D9.12

198

$$\text{Total horizontal component} = 250 \cos 25° + 350 \cos (-35°)$$
$$= 226.6 + 286.7$$
$$= 513.3 \, \text{kN}$$
$$\text{Total vertical component} = 250 \sin 25° + 350 \sin (-35°)$$
$$= 105.7 - 200.8$$
$$= -95.1 \, \text{kN}$$

The resultant vector $\overline{OE}$, drawn from the total horizontal and vertical components, is shown in fig. D9.13.

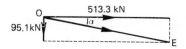

Fig. D9.13

The magnitude of $\overline{OE}$ is calculated using the theorem of Pythagoras:

$$OE^2 = 513.3^2 + 95.1^2$$
$$\therefore \quad OE = 521.9 \, \text{kN}$$

and $\tan \alpha = \dfrac{95.1}{513.3}$

$$\therefore \quad \alpha = \arctan 0.19$$
$$= 10.50°$$

i.e. the resultant force on the rig has magnitude 521.9 kN and acts at an angle of 10.50° south of east.

D9.7 Phasors

In addition to scalar and vector quantities there is a third group of quantities, called *phasors*, which have magnitude and vary with time. Examples of phasors include alternating currents and voltages and the magnetic fields caused by alternating currents. Phasors can be represented in diagrams in the same way as vectors, using a line drawn to scale so that its length corresponds to the magnitude of the phasor. However, the angle made by the line will not represent the direction of the phasor in space (as in a vector diagram) but the *phase angle* associated with the phasor.

Figure D9.14(a) shows the phasors representing two voltage waveforms: $V_1 = 50 \sin x$ and $V_2 = 100 \sin (x + \pi/4)$.

Phasors may be added using methods described in earlier sections on vectors. To find the resultant of the phasors in fig. D9.14(a), we first calculate the total horizontal and vertical components:

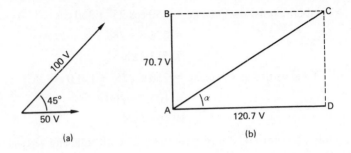

Fig. D9.14 Phasor diagram showing resultant

$$\text{total horizontal component} = 50 + 100 \cos 45°$$
$$= 120.7 \text{ V}$$
$$\text{total vertical component} = 0 + 100 \sin 45°$$
$$= 70.7 \text{ V}$$

Applying the theorem of Pythagoras to these components (see fig. D9.14(b)), the resultant magnitude AC = $\sqrt{(120.7^2 + 70.7^2)}$ = 139.9 V and the phase angle α = arctan $(70.7/120.7)$ = $30.4°$. Thus the resultant phasor is $V = 139.9 \sin (x + 30.4°)$.

Exercise D9

1 Define a vector and state two examples of vector quantities.
2 Draw a vector diagram suggested by the equation $\overline{\text{AD}} = \overline{\text{AB}} + \overline{\text{AC}}$.
3 Explain the words *resolution* and *resultant* as they are applied to vectors.
4 Draw to scale the momentum vector $\overline{\text{OZ}}$ with magnitude 2.7 N s and direction $44°$ measured anticlockwise from the positive x-axis. Measure the components of $\overline{\text{OZ}}$ along the x- and y-axes.
5 Calculate the horizontal and vertical components of vector $\overline{\text{OY}}$ in fig. D9.15, which represents the acceleration of an aircraft.

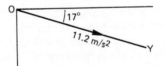

Fig. D9.15

6 Calculate the magnitude and direction (expressed as angle α) of the vector $\overline{\text{OC}}$ from the components at right angles in fig. D9.16.
7 The rotor of a helicopter develops a vertical lift component of x kN and a forward thrust component of $x/6$ kN. Draw a vector diagram to show these components and hence calculate the magnitude of their resultant and the angle which the resultant makes with the horizontal.

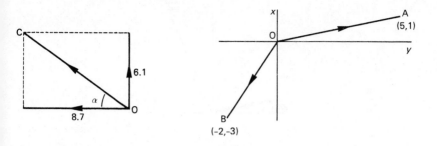

Fig. D9.16   Fig. D9.17

8   Determine the total x- and y-components of the vectors $\overline{OA}$ and $\overline{OB}$ in fig. D9.17.

9   Resolve the three force vectors in fig. D9.18 into component parts at right angles, and calculate the total horizontal and vertical components.

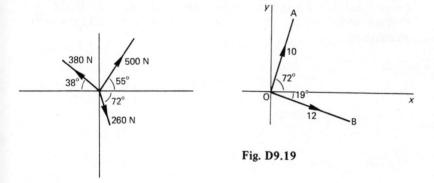

Fig. D9.19

Fig. D9.18

10   Resolve both vectors in fig. D9.19 to find the total x- and y-components. Using these components, calculate the magnitude of the resultant vector and the angle it makes with the x-axis.

11   In fig. D9.20, vector $\overline{OX}$ has magnitude $1000\,\Omega$ and represents the reactance of an inductance in a circuit. Vector $\overline{OR}$ has magnitude $2500\,\Omega$ and represents resistance in the same circuit. Draw the resultant vector $\overline{OZ}$, which denotes the impedance of the circuit, and calculate both its magnitude and the phase angle ϕ which $\overline{OZ}$ makes with $\overline{OR}$.

Fig. D9.20

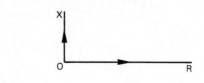

201

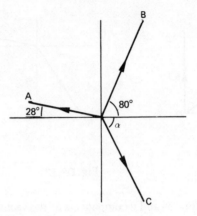

Fig. D9.21

12 Figure D9.21 shows three forces acting at a point. If the magnitudes of forces *A* and *B* are 6 N and 10 N respectively, resolve horizontally and vertically to find the magnitude of force *C* and angle α when the system is in equilibrium.

E   Statistics

E1 Collecting and interpreting data

While it is quite correct to regard *statistics* as a branch of mathematics, the word originally referred to data collected by state departments for governmental purposes. The importance of statistics derives from the assistance it gives in decision-making in so many different areas, ranging from consideration of the economic consequences of population growth to quality control or analysis of the relative effectiveness of insecticides.

E1.1 Types of data
The two types of data we shall distinguish are *discrete* and *continuous*. Figures gathered by counting individual items are known as discrete data. By measurement we get continuous data.

Before we can collect and classify our data we find we have to define precisely what quantity it is we wish to count or measure. For example, in making an analysis of the earnings of a group of workers we would have to define exactly what we meant by 'earnings' (whether they included bonuses, overtime, etc.) and also what we meant by 'workers' (are we including part-timers, directors, etc?).

It is also helpful to make a distinction between the term *variable*, which is commonly used for a quantity subject to continuous change, and the more specialised term *variate* which we shall reserve to denote a quantity able to take any of a number of individual values any number of times. Thus, with equations such as $y = 4x^2$ and $s = 20t + 4.9t^2$, the letters y, x, s, and t represent variables. If we collect information about the number of books (n) sold each day from a certain bookshop, then n is a variate. As it is possible that an equal number of books may be sold on two or more days, we see that a variate has values which can recur. We use the term *frequency* to denote the number of times any particular value occurs. A table giving values of a variate (x) and the corresponding frequencies (f) is called a *frequency distribution*.

E1.2 Arranging data into groups
When the range of values of a variate is very large, it may be subdivided into groups which are usually known as *classes*. If the marks scored in an examination range from 21 to 100, these scores could be grouped together, for example into ten grades each covering 8 possible values or eight grades each covering 10 possible values. If we decide to use grades each covering 10 marks,

then we say that 10 marks is the *class interval*, and it is helpful if the class intervals are of the same size over the entire range of values.

Sometimes unequal class intervals may be used, and for an example we could consider a road-traffic survey to be carried out in an urban area where it was proposed to impose a speed limit of 50 km/h. In this case it might be appropriate to grade the speeds of passing vehicles in the following way: up to 50 km/h; over 50 and up to 65 km/h; over 65 and up to 80 km/h; over 80 and up to 95 km/h; over 95 and up to 110 km/h; over 110 km/h. This would provide larger class intervals for the slowest and for the fastest vehicles, but this would not matter as long as the figures produced conveyed the information required.

It is important to define class boundaries very carefully, especially with continuous data, to avoid any confusion about which class is the right one to which a value should be allocated. One way of doing this is to use careful wording, as in the road-speed example above: an alternative is to express the same class intervals mathematically; for example, 'over 50 and up to 65' can be written as $50 < x \leqslant 65$. We shall see later that it is also important to be precise in defining the central point between class boundaries, since the central value is used subsequently in other calculations and each central value is taken as representative of its respective class.

Consider the following list of the measured heights in metres of 20 students

1.75	1.59	1.70	1.72	1.76
1.52	1.93	1.63	1.85	1.69
1.81	1.77	1.83	1.60	1.81
1.66	1.68	1.90	1.65	1.56

If we group the 20 values into five class intervals of 10 cm, we can cover the range. These class intervals must be clearly defined to avoid overlapping and ambiguity. If we chose 1.50–1.60 m as the *class limits* of the first class interval and 1.60–1.70 m as the second class interval, for example, there would be doubt about whether measurements of exactly 1.60 m went into the first or the second class. To avoid this we let 1.59 m be the *upper class limit* of the first class interval and 1.60 be the *lower class limit* of the second. In the majority of cases we need to record only the frequencies of measurements in each class interval. Tally marks can be used to count how many of the 20 values in the list are in each class, as shown in the table below.

Class interval (m)	Tally marks	Frequency
1.50–1.59	111	3
1.60–1.69	HHl	5
1.70–1.79	HHl 1	6
1.80–1.89	1111	4
1.90–1.99	11	2
	Total	20

Because student height is a continuous variate we may obtain measurements which fall between two class intervals. For example, a height of 1.594 m lies between 1.59 m, the upper class limit of the first class interval, and 1.60, the lower class limit of the next. The point half-way between adjacent class limits is called the *class boundary*. For example the lower and upper class boundaries of the first class interval are 1.495 m and 1.595 m respectively.

For the purposes of drawing diagrams and calculations, we can represent a class interval by its *central value*, which is half-way between the lower and upper class boundaries of that class interval:

$$\text{central value of first class interval} = \frac{1.495 + 1.595}{2} = 1.545 \text{ m}$$

All three measurements in this class interval are now considered to be 1.545 m. This step will obviously introduce errors, since $3 \times 1.545 = 4.635$ m, whereas the true total height in this class interval is 4.67 m. However, with large numbers of variate values the approximation is generally good enough.

Example 1   The speeds of 30 vehicles along a stretch of road were measured to the nearest kilometre per hour, using radar, and are given below. Arrange the data into six equal class intervals in the range 25 to 54 km/h, and draw up a table giving central value and frequencies for each class interval.

28	38	36	31	48	36
50	41	29	37	33	32
32	33	43	45	29	44
44	36	37	40	47	30
51	40	42	39	27	41

With six classes in the range specified, the class interval will be $(54 - 25)/6 \approx 5$ km/h. The class limits of the first interval are therefore 25 and 29 km/h (with class boundaries 24.5 and 29.5 km/h). The central value of this class interval is

$$\frac{24.5 + 29.5}{2} = 27 \text{ km/h}$$

Having worked out all the central values in this way, we draw a table and use a tally method to count the occurrence of speeds in each class, i.e. the frequencies:

Speed (km/h)	Central value (km/h)	Tally marks	Frequency
25–29	27	1111	4
30–34	32	⊓⊔⊔ 1	6
35–39	37	⊓⊔⊔ 11	7
40–44	42	⊓⊔⊔ 111	8
45–49	47	111	3
50–54	52	11	2

When arranging discrete data into appropriate classes, the class boundaries and central values may not be values which could exist. For example, a class of observations of light-bulb failure may have class boundaries of 9.5 and 15.5 bulbs with the central value at 12.5 bulbs. These may appear meaningless, but we can nevertheless use these figures in our calculations. You have probably seen references to families with 2.3 children and 1.4 cars in statistical reports.

Unequal class intervals

It is sometimes useful to use class intervals which are unequal. We may wish to analyse a particular portion of a frequency distribution in more detail or simply to avoid having large numbers of intervals with low frequencies.

Example 2   Tests on the output of 100 car batteries, measured to the nearest ampere hour, gave the following data:

35	38	34	39	40	37	32	36	41	33
29	36	31	36	28	39	43	46	27	31
37	42	37	41	37	41	20	40	38	37
40	23	39	43	45	25	45	48	51	18
17	40	40	31	4	40	39	31	40	39
36	40	27	44	30	63	38	52	55	48
39	47	41	46	38	34	53	43	58	25
41	18	38	51	49	42	57	42	29	53
31	50	44	42	32	47	22	28	33	32
35	41	33	42	38	36	39	43	37	40

Draw a table to show the frequency distribution, using the following class intervals: 0–20, 21–30, 31–35, 36–40, 41–50, and 51–70 A h.

The fact that these class intervals are unequal makes no difference to the way we construct the following table:

Battery output (A h)	Tally marks	Frequency
0–20	ᚋᚋᚋ	5
21–30	ᚋᚋᚋ ᚋᚋᚋ 1	11
31–35	ᚋᚋᚋ ᚋᚋᚋ ᚋᚋᚋ	15
36–40	ᚋᚋᚋ ᚋᚋᚋ ᚋᚋᚋ ᚋᚋᚋ ᚋᚋᚋ ᚋᚋᚋ 111	33
41–50	ᚋᚋᚋ ᚋᚋᚋ ᚋᚋᚋ ᚋᚋᚋ ᚋᚋᚋ 11	27
51–70	ᚋᚋᚋ 1111	9
	Total	100

E1.3 Histograms from grouped data

A histogram is a diagram drawn using information from a frequency table. Values of the measured variate are shown along the horizontal axis, and we draw vertical columns with areas which represent frequencies in each class. When class intervals are equal, we can use the vertical axis to show frequencies.

This is possible because the areas and heights of columns with the same width are in direct proportion. Figure E1.1 shows a histogram drawn from the frequency table of students' heights (see section E1.2). Each column corresponds to one class interval, which is represented by its central value.

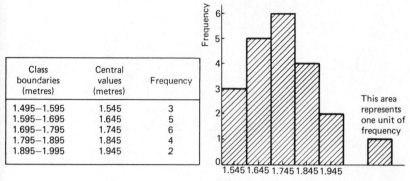

Class boundaries (metres)	Central values (metres)	Frequency
1.495–1.595	1.545	3
1.595–1.695	1.645	5
1.695–1.795	1.745	6
1.795–1.895	1.845	4
1.895–1.995	1.945	2

Fig. E1.1 Frequency table and histogram showing height distribution

Note that in fig. E1.1 the divisions between columns on the histogram are drawn at class boundaries. We could have shown class boundaries on the axis instead of the central values.

Example 1 Draw a histogram from the frequency table below, which was compiled from measurements of the speeds of 30 vehicles (see section E1.2).

Speed (km/h)	*Frequency*
24.5–29.5	4
29.5–34.5	6
34.5–39.5	7
39.5–44.5	8
44.5–49.5	3
49.5–54.5	2

As in the previous example, the class intervals are equal and we can measure frequency along the vertical axis of the histogram (see fig. E1.2).

When class intervals are unequal, we must again remember that it is the *area* of a column on a histogram that is proportional to frequency.

Example 2 The table below resulted from tests on 100 car batteries. Show the frequency distribution on a histogram. (See section E1.2 for the origin of the table.)

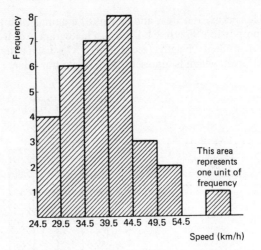

Fig. E1.2 Histogram showing distribution of vehicle speeds

Battery output (Ah)	Frequency
0–20	5
21–30	11
31–35	15
36–40	33
41–50	27
51–70	9
Total	100

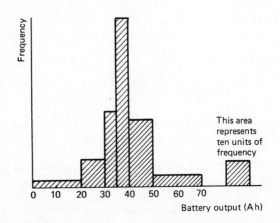

Fig. E1.3 Histogram showing distribution of battery outputs

208

Figure E1.3 shows the histogram drawn from this table. Since the class intervals are not equal in this case, we do not use the vertical axis to denote frequency values. It is especially important in such cases to indicate the scale on your histogram.

E1.4 Relative frequency

The relative frequency with which any value occurs is given by the frequency of that item divided by the total of the frequencies of all the items.

For a grouped distribution, the relative frequency of any class is given by the frequency of that class divided by the total of the frequencies of all the classes. Relative frequencies are usually expressed as percentages. Changing the frequencies on the vertical scale of a histogram into percentages changes the diagram to a relative-frequency histogram.

Example A group of companies analyses its pension scheme for female employees and the following table shows the age distribution of all those who are at present receiving benefit.

Age group	60–64	65–69	70–74	75–79	80–84	85–89	90–95
Frequency %	15	27	24	18	10	5	1

The relative-frequency histogram is shown in fig. E1.4.

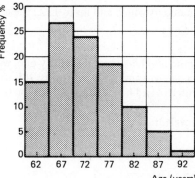

Fig. E1.4 Relative-frequency histogram showing age distribution of female pensioners

When constructing histograms it is advisable to choose scales such that the height-to-width ratio is between 1:2 and 1:1, since this helps to produce a histogram which looks well-balanced. Note that every statistical diagram should be given a title, and the various sections of a chart or the axes of the histogram or statistical curve should be clearly labelled.

E1.5 Types of average

The arithmetic mean

The arithmetic mean is often referred to simply as 'the mean', although this is not recommended. It is the type of average most widely understood, being used even in sport (e.g. batting averages), and is obtained by adding together all the values of the items in the distribution and dividing the total by the number of items. This is conveniently expressed as

$$\bar{x} = \frac{1}{n} \Sigma x$$

where $\bar{x}$ is the arithmetic mean, n is the number of items in the distribution, Σ is a mathematical symbol meaning 'the sum of all the values of', and x is the variate.

For a grouped distribution each value of the variate x must be multiplied by the corresponding frequency f, and in this case the arithmetic mean is given by

$$\bar{x} = \frac{1}{n} \Sigma f x$$

Example 1   Find the average of the following numbers: 9, 9, 9, 9, 9, 9, 12, 12, 12, 14, 14, 14.

i) Here we have 12 numbers which add up to 132. The arithmetic mean is

$$\bar{x} = \frac{1}{12} (132) = 11$$

ii) Instead of adding the numbers together one by one, we could total them as six nines plus three twelves and three fourteens. It is easy to see that this gives the same total of 132 and hence the same arithmetic mean of $\bar{x} = 11$.

Method (i) is equivalent to $\bar{x} = \frac{1}{n} \Sigma x$

where $\Sigma x = 9 + 9 + 9 + 9 + 9 + 9 + 12 + 12 + 12 + 14 + 14 + 14$

Method (ii) is equivalent to $\bar{x} = \frac{1}{n} \Sigma f x$

where $\Sigma f x = (6 \times 9) + (3 \times 12) + (3 \times 14)$

The calculation can be tabulated as follows:

x	f	fx
9	6	54
12	3	36
14	3	42
	12	⟩132
	$\bar{x} =$	11

Such tabulation is unnecessary in a simple case such as this but is helpful when handling a lot of data.

The median

The median is the central value of a distribution. To find the median, the complete distribution must be arranged in order of size, that is *ranked*; then, with an odd number of items, the central value will be the median. With an even number of items the median is the average of the central pair.

Example 2   Find the median of the set of numbers 2, 5, 7, 9, 12, 13, 17.

In this case there is an odd number of items (seven) and the central value 9 is the median.

Example 3   Find the median of the set of numbers 9, 9, 9, 9, 9, 9, 12, 12, 12, 14, 14, 14.

In this case, with an even number of items (twelve), the central pair is 9 and 12 and the median is thus $\frac{1}{2}(9 + 12) = 10.5$.

Finding the median from ungrouped data

The chief virtue of the median is that it gives a typical central value which is unaffected by extremes. This can be very helpful if information is incomplete at either end of the distribution. For example, an 'average salary' determination may be hampered by complications over sick-pay benefit at one end of the scale of salary payments and by a refusal to disclose the salary of the highest paid at the other end, yet calculation of the median is still possible.

Example 4   Information is collected on a sample of 25 standard 60 watt electric light bulbs to determine an average for the length of time such a bulb can be expected to provide illumination. Find the median given that one bulb had a broken filament when purchased, three were still operating when the testing was concluded, and the times the other 21 lasted under continuous use were 36, 40, 39, 43, 39, 37, 41, 38, 37, 38, 44, 41, 45, 39, 46, 40, 46, 45, 42, 41, 43 days.

The first task is to arrange the times in rank order, as follows: $<$36, 36, 37, 37, 38, 38, 39, 39, 39, 40, 40, 41, 41, 41, 42, 43, 43, 44, 45, 45, 46, 46, $>$46, $>$46, $>$46. Of 25 values, the central one will be the thirteenth, which in this case is 41; i.e. the median is 41 days.

Quartiles and percentiles

We have seen that the median of a set of ranked data is the central value or the average of the central pair of values and that it divides the distribution into two equal parts. The values which further divide these halves into two equal parts are called quartiles. The lower quartile Q_1 divides the data below the median into two equal parts and the upper quartile Q_2 divides the data above the median in the same way.

Example 5 Find the lower and upper quartiles for the following eight numbers: 6, 3, 11, 2, 7, 5, 13, 10.

First we rank the numbers, to give

$$2, 3, 5, 6, 7, 10, 11, 13$$

The lower quartile Q_1 is the value half-way between the second and third values:

$$Q_1 = \frac{3 + 5}{2} = 4$$

The upper quartile Q_2 is half-way between the sixth and seventh values:

$$Q_2 = \frac{10 + 11}{2} = 10.5$$

Percentiles divide a distribution into one hundred equal parts, so that the lower quartile, median, and upper quartile correspond to the 25th, 50th, and 75th percentiles respectively.

The mode
The mode is the most frequently occurring value in a distribution.

Example 6 Find the mode of the set of numbers 9, 9, 9, 9, 9, 9, 12, 12, 12, 14, 14, 14.

The value 9 occurs with the greatest frequency, and thus the mode in this case is 9.

Choice of average
Each of these averages possesses certain advantages and they are used in different circumstances, the aim in each case being to select the particular average which is most suitable to be taken as representative of the whole distribution.

It should be noted that the arithmetic mean uses every value in the distribution but is unduly affected by the existence of extreme items and may not coincide with any original values (e.g. 'each family had an average of $2\frac{1}{4}$ children'!) The arithmetic mean has the advantage of being well known and, furthermore, the total of all the items in the distribution may be regained from it, since

$$\Sigma x = n\bar{x}$$

The arithmetic mean is used in the analysis of experimental results when they are reasonably consistent, and is the basis of many further statistical calculations.

With discrete data it is convenient to use the mode when the type of average required should be one of the actually occurring values (e.g. 'the average Englishman wears a shirt with collar size 39 cm).

With continuous data, the mode corresponds to the peak of a frequency curve (see page 229).

Where the range can be stipulated, a histogram can be drawn and the values of arithmetic mean, median, and mode determined. For a distribution which is symmetrical with a central peak, the arithmetic mean, the median, and the mode should coincide. When the peak lies to one side of the mean, the distribution is said to be *skewed* (fig. E1.5). For such a skewed distribution, the value of the median will be between the arithmetic mean and the mode.

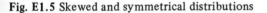

Positive skew   Negative skew   Symmetrical

Fig. E1.5 Skewed and symmetrical distributions

Example 7 In a survey to determine the length of time spent by individual customers buying groceries in a certain supermarket, times were recorded to the nearest minute for 1000 customers over the course of a week:

Time (minutes)	1–10	11–20	21–30	31–40	41–50	51–60
No. of customers	171	387	237	105	68	32

Illustrate this distribution by a histogram, and give the values of the range, arithmetic mean, and mode.

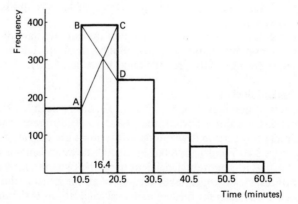

Fig. E1.6 Histogram showing shopping times

213

The histogram is shown in fig. E1.6.

The range is one hour.

The class boundaries are 0.5, 10.5, 20.5, etc. to 60.5 minutes.

The central values are thus 5.5, 15.5, 25.5, etc. to 55.5 minutes.

x	f	xf
5.5	171	940.5
15.5	387	5 998.5
25.5	237	6 043.5
35.5	105	3 727.5
45.5	68	3 094.0
55.5	32	1 776.0
	1000	)21 580
	$\bar{x}$ =	21.6

i.e. the arithmetic mean is 21.6 minutes.

The modal class is the second one (11–20), which has the greatest frequency. A method of estimating a value for the mode is shown on the histogram in fig. E1.6. The time corresponding to the point at which lines AC and BD cross is used. This gives the value 16.4 minutes for the mode.

Range

Example 4 in this section illustrates the difficulty sometimes experienced in defining the *range* of a distribution. Normally we would define range as being the difference between the largest and the smallest values of the variate, and this concept is a valuable one where these values are known, although if there is a single extreme value markedly different from the rest this fact should be noted separately. One illustration of this could be a stamp collection of ordinary inexpensive stamps in which was discovered a rare variety worth over £2000. In such a case it would hardly convey a fair impression to state that the collection contained stamps ranging in value from 1 p to £2000! It might be more realistic to say that the range of stamp values was up to £3 but for the one exception noted.

The range of a distribution is the simplest measure we have of the *spread* or *dispersion* of a distribution and is helpful when comparing one distribution with another. (See also inter-quartile range in section E2.3.)

E1.6 Standard deviation

The two sets of numbers 3, 4, 5, 6, 7 and 1, 3, 5, 7, 9 each have the same arithmetic mean, but it is obvious that the *spread* is different in the two cases. In this simple comparison it would be sufficient to use the *range* as a measure of dispersion, 3 to 7 and 1 to 9, but in general the range is not a very satisfactory guide as it is entirely dependent upon the two extreme values. The measure of dispersion which takes account of all the items in the distribution and is used as the basis for further statistical calculations is the *standard deviation*.

Basically, the standard deviation is the root-mean-square value of deviations from the arithmetic mean. To calculate it, we first find by how much each value of the variate x differs from the mean, then square these deviations from the mean, then calculate the mean of the squares of deviations, and finally take the square root.

The symbol σ ('*sigma*') is used for standard deviation, and its mathematical determination may be expressed as

$$\sigma = \sqrt{\frac{1}{n}\Sigma(x-\bar{x})^2}$$

where n is the total number of items and $\bar{x}$ is the arithmetic mean.

For a grouped distribution where for each class x is the central value and f the frequency, the standard deviation is given by

$$\sigma = \sqrt{\frac{1}{n}\Sigma f(x-\bar{x})^2}$$

The square of the standard deviation (σ^2) is also used as a measure of the spread of a distribution and is known as the *variance*.

When comparing two sets of samples, it is convenient to express the standard deviation as a percentage of the mean to give the *coefficient of variation*.

For the two sets of numbers 3, 4, 5, 6, 7 and 1, 3, 5, 7, 9, the calculation of the standard deviations may conveniently be set out as follows:

Set 1			Set 2		
Variate (x)	*Deviation* $(x-\bar{x})$	*(Deviation)*^{2} $(x-\bar{x})^2$	*Variate* (x)	*Deviation* $(x-\bar{x})$	*(Deviation)*^{2} $(x-\bar{x})^2$
3	−2	4	1	−4	16
4	−1	1	3	−2	4
5	0	0	5	0	0
6	1	1	7	2	4
7	2	4	9	4	16
5)25		5)10	5)25		5)40
$\bar{x}=5$		2	$\bar{x}=5$		8
	$\sigma=\sqrt{2}$			$\sigma=\sqrt{8}=2\sqrt{2}$	

(5 items for Set 1, 5 items for Set 2)

Thus the standard deviation for the second set is double that for the first.

It is advisable to adopt a suitable tabular form for setting out the calculation of the mean and the standard deviation, as illustrated by the following examples.

Example 1 The labour expenditure (man-hours per house) of twenty skilled bricklayers is given below. Calculate the arithmetic mean and the standard deviation.

| 603 | 625 | 559 | 703 | 549 | 658 | 618 | 589 | 638 | 597 |
| 512 | 645 | 544 | 611 | 565 | 583 | 678 | 574 | 667 | 482 |

No. of hours (x)	Deviation $(x - \bar{x})$		$(Deviation)^2$ $(x - \bar{x})^2$
603	+3		9
625	+25		625
559		−41	1 681
703	+103		10 609
549		−51	2 601
658	+58		3 364
618	+18		324
589		−11	121
638	+38		1 444
597		−3	9
512		−88	7 744
645	+45		2 025
544		−56	3 136
611	+11		121
565		−35	1 225
583		−17	289
678	+78		6 084
574		−26	676
667	+67		4 489
482		−118	13 924
20)12 000	+446 −446		20)60 500
$\bar{x}$ = 600 h	Check		3 025

$$\sigma = \sqrt{3\,025} = 55 \text{ h}$$

i.e. the arithmetic mean is 600 hours and the standard deviation is 55 hours.

Some calculators will give the arithmetic mean and standard deviation of a list of numbers automatically if the numbers are entered one at a time.

Example 2 The distribution of the ages of 64 people working in a factory is given below. Calculate the mean age and the standard deviation.

Age (years)	16−20	21−25	26−30	31−35	36−40	41−45	46−50
No. with this age	7	14	17	12	8	5	1

Here we have a grouped distribution, and it is convenient to regard the group 16−20 as though it were the single central value 18. It is these central values that we shall in fact use for our variates. This may introduce a slight inaccuracy but this is unavoidable unless the 64 original values are known, and even then

216

it is such a lengthy and tedious operation to deal separately with each one that it is very seldom worthwhile.

To find the standard deviation using the formula given earlier, it is necessary to calculate the mean value first. All the steps in the computation of the mean and standard deviation are shown in the table below.

Age (years)	Central value (x)	Frequency (f)	fx	$x - \bar{x}$	$f(x - \bar{x})^2$
16–20	18	7	126	−11.5	925.8
21–25	23	14	322	−6.5	591.5
26–30	28	17	476	−1.5	38.3
31–35	33	12	396	3.5	147.0
36–40	38	8	304	8.5	578.0
41–45	43	5	215	13.5	911.3
46–50	48	1	48	18.5	342.3
Totals		64	1887		3534.2

$$\bar{x} = \frac{\Sigma fx}{n} = \frac{1887}{64} = 29.5 \text{ years}$$

The last two columns in the table are completed using $\bar{x} = 29.5$.

$$\sigma = \sqrt{\frac{\Sigma f(x - x)^2}{n}} = \sqrt{\frac{3534.2}{64}} = 7.4 \text{ years}$$

i.e. the mean age is 29.5 years and the standard deviation is 7.4 years.

These calculations can be performed quickly using a calculator. The mean value can be computed using a continuous series of calculator operations (i.e. $[(18 \times 7) + (23 \times 14) + \ldots + 48] \div 64$).

There is a second formula for standard deviation which can easily be derived from the one we have used:

$$\sigma = \sqrt{\frac{\Sigma fx^2}{n} - \left(\frac{\Sigma fx}{n}\right)^2}$$

This version is more convenient to use and is usually more accurate, since it does not involve repeated use of $\bar{x}$: any rounding of the arithmetic mean will lead to slight errors in the earlier formula for standard deviation.

Using data from the previous example, we can draw up the table shown on page 218 and calculate σ as follows.

The standard deviation is given by

$$\sigma = \sqrt{\frac{\Sigma fx^2}{n} - \left(\frac{\Sigma fx}{n}\right)^2} = \sqrt{\frac{59\,171}{64} - \left(\frac{1887}{64}\right)^2} = 7.4 \text{ years}$$

Using this method with a calculator which has memory facilities, we can compute the standard deviation from the table of x and f values very quickly.

Central value (x)	Frequency (f)	fx	fx²
18	7	126	2 268
23	14	322	7 406
28	17	476	13 328
33	12	396	13 068
38	8	304	11 552
43	5	215	9 245
48	1	48	2 304
Totals	64	1887	59 171

By planning the sequence of calculator operations carefully, we can work through the table accumulating values of Σfx and Σfx^2 to be used in the formula.

When very large or very small numbers are involved in the calculation of arithmetic mean and standard deviation, it is better to work in standard form throughout. Many scientific calculators will operate in this way.

Exercise E1

1 Which of the following examples of statistical data are discrete?

a) Production of television sets by a manufacturer
b) The heights of several piles of pennies
c) Current flowing through a wire
d) Thickness of chromium plate
e) Tensile strength of test specimens
f) Yearly industrial-accident figures.

2 Toss a coin twenty times and record the frequency of heads and tails on a table, using tally marks.

3 Fifty resistors were checked to the nearest ohm. Make a frequency table from the following results, using tally marks for each of the values occurring.

100	99	97	99	103	97	100	101	101	102
99	100	98	100	100	99	103	102	99	102
100	98	99	99	100	100	102	100	101	100
99	99	102	100	100	101	101	99	101	98
100	102	99	99	101	99	101	101	97	100

4 Samples of 20 valve springs were taken from a production line at regular intervals and tested. The numbers of rejects in each of 50 samples are given below. Make a frequency table of the numbers rejected.

0	0	0	0	0	1	1	2	1	3
1	0	1	0	4	2	0	1	0	1
0	1	2	0	0	0	0	1	0	0
2	0	1	3	2	0	2	0	1	0
0	3	0	1	1	1	1	0	2	1

5 Define the *range* of a set of data.

6 Arrange in the form of a frequency table the following data taken from fuel economy tests (in kilometres per litre) on 30 engines. Use five equal class intervals from 16.5–16.9 km/l to 18.5–18.9 km/l.

17.2	18.6	17.7	17.3	16.8	17.7
18.1	17.6	18.2	17.7	17.3	17.2
17.5	17.4	17.9	18.0	17.8	17.6
17.7	17.2	17.1	18.3	18.1	18.0
16.5	16.9	16.8	17.6	17.3	16.6

7 Two adjacent intervals in a frequency distribution have class limits 200.5–201.4 mm and 201.5–202.4 mm. What are the class boundaries of these intervals?

8 If five equal class intervals in a frequency distribution, drawn from current measurements, have central values 0.225 A, 0.285 A, 0.345 A, 0.405 A, and 0.465 A, what are the class boundaries of the highest interval?

9 Draw a fully labelled histogram using the following frequency table of 70 motor commutator diameters measured during quality-control tests. Use your diagram to estimate the mode.

Class boundaries (mm)	Frequency
32.045–32.065	3
32.065–32.085	7
32.085–32.105	11
32.105–32.125	22
32.125–32.145	14
32.145–32.165	9
32.165–32.185	4

10 Draw a histogram to represent the following table of sales figures for motor cycles over a range of engine sizes.

Class interval (cm³)	Sales
Up to 100	8 000
101–250	15 000
251–350	7 000
351–500	4 000
501–1000	2 000

11 The following table gives the marks awarded to 100 students in an examination:

x	1–10	11–20	21–30	31–40	41–50	51–60	61–70	71–80	81–90	91–100
f	1	4	8	17	24	21	14	6	3	2

Draw the histogram of this distribution.

12 Arrange the following nine numbers into ranking order and find the median: 17, 10, 14, 23, 20, 24, 9, 15, 7.

13 Define the lower and upper quartiles of a distribution and find them for the following list of numbers: 52, 36, 42, 66, 69, 31, 40, 45, 57, 49, 63, 51.

14 In the same examination, 22 students in one class gained an average of 46 marks, while 28 students in another class had an average of 51 marks. Calculate the overall average for the two classes together.

15 In absorption tests on 200 bricks, the following figures were obtained:

% absorption	6	7	8	9	10	11	12	13	14	15
Frequency	1	5	13	31	51	47	33	14	4	1

Calculate the arithmetic mean and the standard deviation for these results.

16 The number of hours worked in a certain week by each of 100 employees is recorded as follows:

Hours per man	40	41	42	43	44	45	46	47	48
No. of men	3	4	8	12	17	21	19	11	5

Construct a histogram for these results. Calculate the values of the arithmetic mean and the standard deviation.

17 The table shows the wages paid by a firm to 1000 of its employees:

Wages (£)	44→48	→52	→56	→60	→64	→68	→72	→76	→80
No. of employees	53	73	140	190	209	170	86	57	22

Calculate the mean wage and the standard deviation from it.

18 Calculate the mean and the standard deviation for the following set of test results:

x	6	8	10	12	14	16	18
f	4	55	244	394	244	55	4

Deduce the mean and standard deviation of a second set of results for which the distribution is

x	7	9	11	13	15	17	19
f	4	55	244	394	244	55	4

19 The following table shows the results of crushing tests on a number of concrete cubes tested in the laboratory. Draw a histogram for these results and read off the mode. Calculate the value of the mean and the standard deviation.

Crushing stress (MPa)	20	22	24	26	30	32	34
No. of cubes failing at this stress	2	7	20	19	4	3	1

E2 Frequency-distribution curves

E2.1 Cumulative frequency

Consider the table below, which shows the distribution of the individual distances travelled to work each day by 100 people in an office. The range of values 1–30 km has been broken down into six class intervals of 5 km. Frequencies in each class interval are recorded. (Note that in the table distances have been given to the nearest kilometre.)

Distance travelled to work (km)	Frequency
1–5	8
6–10	17
11–15	33
16–20	22
21–25	15
26–30	5
Total	100

If we wish to know how many of the 100 people travel 15 km or less to work, we must add the frequencies in the first three classes. The total or *cumulative* frequency is 8 + 17 + 33 = 58. We can extend the table to show cumulative frequencies for all the classes as follows:

Distance (km)	Frequency	Cumulative frequency
1–5	8	8
6–10	17	25
11–15	33	58
16–20	22	80
21–25	15	95
26–30	5	100

Note that the final cumulative frequency value must always be the same as the size of the population or sample from which the statistical data was taken.

Example 1   Add a column of cumulative-frequency values to the table below (which shows the distribution of 200 quarterly electricity bills) and find the percentage of the sample paying £26.50 or less.

Electricity bill (£)	Frequency
2–6	8
7–11	19
12–16	25
17–21	41
22–26	47
27–31	32
32–36	19
37–41	9
Total	200

We find the cumulative-frequency values for each interval in turn by adding the frequency in the interval to the sum of those in previous intervals. The cumulative frequency in the second interval is therefore $19 + 8 = 27$. The complete table is as follows:

Electricity bill (£)	Frequency	Cumulative frequency
2–6	8	8
7–11	19	27
12–16	25	52
17–21	41	93
22–26	47	140
27–31	32	172
32–36	19	191
37–41	9	200

From the table, the cumulative frequency of 140 for the class £22–£26 is the number of bills which were less than £26.50 (the class boundary). In the sample, the percentage of bills of £26.50 or less is therefore

$$\frac{140}{200} \times 100 = 70\%$$

Using percentages

It is sometimes preferable to express cumulative frequencies as percentages. In an earlier example, dealing with distances travelled to work, in which 100 values were involved, the cumulative frequencies were automatically expressed as percentages. When sample sizes other than 100 are involved, we can perform a simple calculation to obtain percentages. For example, for a cumulative frequency of 36 for a particular class in a distribution of 150 values,

$$\text{percentage cumulative frequency} = \frac{36}{150} \times 100 = 24\%$$

Example 2   The table below gives the distribution of examination results for 335 students. Write down the percentage cumulative frequencies.

Examination results %	0–9	10–19	20–29	30–39	40–49	50–59	60–69	70–79	80–89	90–99
Frequency	7	18	31	55	79	64	40	22	13	6

We first produce a column of cumulative-frequency values by adding each frequency in turn to the sum of all previous frequencies. For the 10–19% group this figure is $18 + 7 = 25$, and for the 20–29% group it is $25 + 31 = 56$. Percentages are found by performing the following calculation for each cumulative-frequency value:

$$\text{percentage cumulative frequency} = \frac{\text{cumulative frequency}}{\text{total number in sample}} \times 100$$

For the 10–19% interval this figure is

$$\frac{25}{335} \times 100 = 7.46\%$$

The completed table is shown below.

Examination marks (%)	Frequency	Cumulative frequency	%
0–9	7	7	2.09
10–19	18	25	7.46
20–29	31	56	16.72
30–39	55	111	33.13
40–49	79	190	56.72
50–59	64	254	75.82
60–69	40	294	87.76
70–79	22	316	94.33
80–89	13	329	98.21
90–99	6	335	100.00

E2.2 The cumulative-frequency curve

We saw in section E1.3 that a frequency distribution can be represented diagrammatically by a histogram. With cumulative frequencies, a *cumulative-frequency curve* can be drawn by plotting values of the variate (or upper class boundaries in the case of grouped data) against cumulative frequencies.

In the table on page 224, the numbers of defective die-cast components in each of 100 batches taken from a production line are shown. Plotting these values with cumulative frequencies and numbers of defectives measured along vertical and horizontal axes respectively gives the curve in fig. E2.1.

Defective components per batch	Frequency	Cumulative frequency
0	1	1
1	4	5
2	12	17
3	23	40
4	28	68
5	18	86
6	10	96
7	3	99
8	1	100

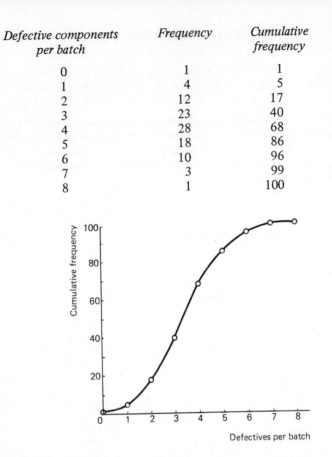

Fig. E2.1 A cumulative-frequency curve

Note the characteristic shape of the cumulative-frequency curve. This curve can be used to obtain statistical information about a distribution not immediately available from the frequency table.

Grouped data
To draw a cumulative-frequency curve when class intervals are given in a frequency table, we must plot cumulative frequencies against *upper class boundaries*. In this way we are including all values up to the beginning of a new class interval with each point we plot. It is also conventional to begin the horizontal scale at the lower class boundary of the first class.

Example Measurements of the daily steel output from a hot-rolling plant over a period of 60 days are shown in the following frequency table. Draw the distribution curve and estimate the number of days in which less than 77.5 t of steel was produced.

Steel output (tonnes)	70.5– 72.5	72.5– 74.5	74.5– 76.5	76.5– 78.5	78.5– 80.5	80.5– 82.5	82.5– 84.5
Frequency	2	7	14	19	11	6	1

We write down the cumulative frequencies for the class intervals in the table as follows:

Cumulative frequency	2	9	23	42	53	59	60

These are plotted against the corresponding upper class boundaries to give the curve in fig. E2.2.

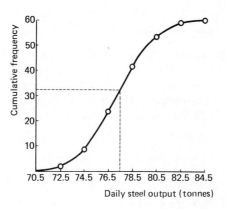

Fig. E2.2

By drawing a vertical line from the horizontal axis at 77.5 t to the curve and across to the cumulative-frequency axis (see the broken line in fig. E2.2), we see that steel output was less than 77.5 t on about 33 of the 60 days.

E2.3 Median and quartiles

We have seen in section E1.5 that the *median* of a distribution of ungrouped variables can be determined by arranging the data in order of size and selecting the central value. (In the case of an even number of values we take the average of the central pair.)

A cumulative-frequency table can be used to determine the median when our data is grouped into class intervals. Consider the cumulative-frequency table in section E2.2, involving data from a steel-rolling plant, which is reproduced below:

225

Steel output (tonnes)	70.5– 72.5	72.5– 74.5	74.5– 76.5	76.5– 78.5	78.5– 80.5	80.5– 82.5	82.5– 84.5
Cumulative frequency	2	9	23	42	53	59	60

As there are an even number of observations involved (60 days), we take the central value to be 30.5, half-way between the 30th and 31st. We can see from the table that this value is in the interval 76.5–78.5 t. There are 23 observations in the range 70.5 to 76.5 t and a further 19 in the interval containing the median. We make the assumption that steel output increases *linearly* over these 19 values and calculate the median by adding a fraction of the class interval (2 t) to the lower class boundary (76.5 t) as follows:

$$\text{observation } 30.5 = 76.5 + \left(\frac{30.5 - 23}{19} \times 2 \right)$$

$$= 76.5 + 0.79 \text{ t}$$

$$\therefore \qquad \text{median} = 77.29 \text{ t}$$

Note that in this example the median is close to the arithmetic mean of the distribution, 77.23 t (estimated by summing the products of central value and frequency for each class interval). This is not always the case, and there are occasions when the median may provide a more representative average value for a distribution than the arithmetic mean.

Example   Calculate the median life of wheel bearings from the following cumulative-frequency table, compiled during tests on 150 bearings.

Bearing life (1000's of hours)	9.5– 12.5	12.5– 15.5	15.5– 18.5	18.5– 21.5	21.5– 24.5	24.5– 27.5	27.5– 30.5	30.5– 33.5
Cumulative frequency	4	15	36	79	121	139	147	150

Since there are 150 bearings (an even number), we take the median life to be half-way between the 75th and 76th observations. Looking at the table, we can see that this value lies in the class interval 18 500 h–21 500 h. There are 43 observations in this interval (found by subtracting 36 from 79) and 36 in the three preceding intervals. We calculate the median by adding a fraction of a 3000 h interval to the lower class boundary (18 500 h) as follows:

$$\text{observation } 75.5 = 18\,500 + \left(\frac{75.5 - 36}{43} \times 3000 \right)$$

$$= 18\,500 + 2755 \text{ h}$$

$$\therefore \quad \text{median bearing life} = 21\,260 \text{ h}$$

Median from the distribution curve
The median can easily be found from the distribution curve by drawing a horizontal line cutting the vertical scale in half and measuring the distance

along the horizontal axis at which the line cuts the curve. The curve in fig. E2.3 is drawn from the wheel bearing-life data used in the previous example. To find the median graphically, we divide the cumulative-frequency scale into two equal parts and observe that this line (drawn from A) meets the curve at a point M which is 21 300 h along the horizontal axis, confirming our calculated median.

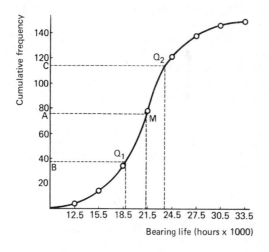

Fig. E2.3 Median and quartiles from a distribution curve

The median, as we have said, divides the distribution into two equal parts. (Note that a line drawn vertically on a histogram at the median will cut the total *area* in half.) If these halves are divided once more (see the lines at B and C in fig. E2.3), we obtain the lower and upper *quartiles*. They are the distances, measured horizontally, to Q_1 and Q_2 respectively. If measured accurately on a well-drawn curve we would find that Q_1 = 18 640 h and Q_2 = 23 770 h. Alternatively we could calculate their values by a method similar to the one we used to find the median, by obtaining the 38th and 113th observations from the cumulative-frequency table.

The semi-interquartile range is sometimes used as a measure of *dispersion* (or spread — see section E2.5). In our example

semi-interquartile range = $\frac{1}{2}(Q_2 - Q_1)$ = 2565 h

If we use percentage cumulative frequencies (see section E2.1) to draw the curve, the median is at the 50th *percentile*, with the lower and upper quartiles at the 25th and 75th percentiles.

227

E2.4 Drawing frequency-distribution curves

The frequency polygon

This is an alternative to the histogram, but is only to be preferred when one distribution is superimposed upon another for purposes of comparison.

With the frequency polygon, the intervals along the axis must be equal, and an additional value (of zero frequency) should be included at each end of the given range of the distribution, so that the area enclosed by the polygon may represent the total frequency.

The frequency polygon is sometimes superimposed on the histogram simply by joining the tops of the columns, but there is no point in doing this and the practice is not recommended.

The following table gives the results of a survey of the number of electricity power points in 75 houses chosen at random:

Number of points	2	3	4	5	6	7	8	9	10	11	12
Frequency	1	3	7	15	20	13	8	4	2	1	1

The frequency polygon drawn using these values is shown in fig. E2.4.

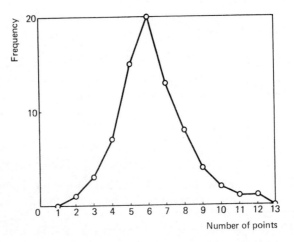

Fig. E2.4 Frequency polygon

The frequency curve

When the figures in the frequency-distribution table represent only a sample, it is reasonable to assume that if more such samples were taken the figures would be similar but not exactly the same. The net effect of combining the frequency polygons for a whole set of samples would be to smooth out the outline, and the result would be a *frequency curve* like the one shown in fig. E2.5.

228

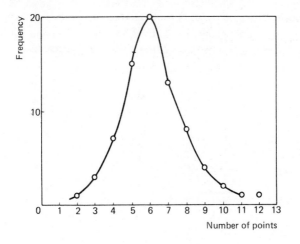

Fig. E2.5 Frequency curve

E2.5 Types of frequency curve

A frequency curve provides a useful pictorial impression of the nature of a distribution. Characteristics of the curve which may have practical value are *symmetry* and *dispersion*.

Symmetry

The curve in fig. E2.6(a) is symmetrical, and we know that the arithmetic mean, mode, and median will all have the same value for this distribution. The frequency curve in fig. E2.6(b) is asymmetrical or *skewed*. We say this curve has *positive skew*. The mode is smaller than the arithmetic mean, with the median lying somewhere between them. The median bisects the area below the curve.

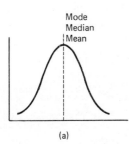

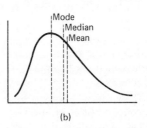

Fig. E2.6

229

Dispersion

The frequency curves shown in fig. E2.7 both have the same arithmetic mean, median, and mode. However, they have different dispersions as measured by their *range* or *standard deviation* (see section E1.6).

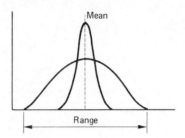

Fig. E2.7

E2.6 The normal curve

In many practical cases when large numbers of observations of a variable are made, the frequency curve we draw from them resembles closely a standard curve we call the *normal curve*. This curve (see fig. E2.8) is widely used in statistical work and tables of areas bounded by the curve and various ordinates are available (see Table E2.1).

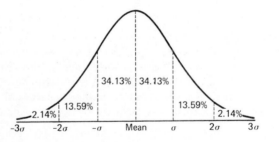

Fig. E2.8 Normal distribution curve

The equation of the normal curve can be expressed in various forms, but the exponential functions involved are beyond the scope of this book. However, noting the following facts is helpful when sketching a normal curve (fig. E2.9):

i) The curve is symmetrical about the mean, and is often shown with the y-axis at this central value.

ii) The curve never meets the x-axis at an angle, but approaches closer and closer to the x-axis at either end.

230

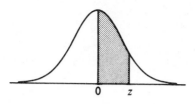

Z	0.00	0.01	0.02	0.03	0.04	0.05	0.06	0.07	0.08	0.09
0.0	0.000	0.004	0.008	0.012	0.016	0.020	0.024	0.028	0.032	0.036
0.1	0.040	0.044	0.048	0.052	0.056	0.060	0.064	0.067	0.071	0.075
0.2	0.079	0.083	0.087	0.091	0.095	0.099	0.103	0.106	0.110	0.114
0.3	0.118	0.122	0.126	0.129	0.133	0.137	0.141	0.144	0.148	0.152
0.4	0.155	0.159	0.163	0.166	0.170	0.174	0.177	0.181	0.184	0.188
0.5	0.191	0.195	0.198	0.202	0.205	0.209	0.212	0.216	0.219	0.222
0.6	0.226	0.229	0.232	0.236	0.239	0.242	0.245	0.249	0.252	0.255
0.7	0.258	0.261	0.264	0.267	0.270	0.273	0.276	0.279	0.282	0.285
0.8	0.288	0.291	0.294	0.297	0.300	0.302	0.305	0.308	0.311	0.313
0.9	0.316	0.319	0.321	0.324	0.326	0.329	0.331	0.334	0.336	0.339
1.0	0.341	0.344	0.346	0.348	0.351	0.353	0.355	0.358	0.360	0.362
1.1	0.364	0.367	0.369	0.371	0.373	0.375	0.377	0.379	0.381	0.383
1.2	0.385	0.387	0.389	0.391	0.393	0.394	0.396	0.398	0.400	0.401
1.3	0.403	0.405	0.407	0.408	0.410	0.411	0.413	0.415	0.416	0.418
1.4	0.419	0.421	0.422	0.424	0.425	0.426	0.428	0.429	0.431	0.432
1.5 0	0.433	0.434	0.436	0.437	0.438	0.439	0.441	0.442	0.443	0.444
1.6	0.445	0.446	0.447	0.448	0.449	0.451	0.452	0.453	0.454	0.454
1.7	0.455	0.456	0.457	0.458	0.459	0.460	0.461	0.462	0.462	0.463
1.8	0.464	0.465	0.466	0.466	0.467	0.468	0.469	0.469	0.470	0.471
1.9	0.471	0.472	0.473	0.473	0.474	0.474	0.475	0.476	0.476	0.477
2.0	0.477	0.478	0.478	0.479	0.479	0.480	0.480	0.481	0.481	0.482
2.1	0.482	0.483	0.483	0.483	0.484	0.484	0.485	0.485	0.485	0.486
2.2	0.486	0.486	0.487	0.487	0.487	0.488	0.488	0.488	0.489	0.489
2.3	0.489	0.490	0.490	0.490	0.490	0.491	0.491	0.491	0.491	0.492
2.4	0.492	0.492	0.492	0.493	0.492	0.493	0.493	0.493	0.493	0.494
2.5	0.494	0.494	0.494	0.494	0.494	0.495	0.495	0.495	0.495	0.495
2.6	0.495	0.495	0.496	0.496	0.496	0.496	0.496	0.496	0.496	0.496
2.7	0.497	0.497	0.497	0.497	0.497	0.497	0.497	0.497	0.497	0.497
2.8	0.497	0.498	0.498	0.498	0.498	0.498	0.498	0.498	0.498	0.498
2.9	0.498	0.498	0.498	0.498	0.498	0.498	0.499	0.499	0.499	0.499
3.0	0.499	0.499	0.499	0.499	0.499	0.499	0.499	0.499	0.499	0.499
3.1	0.499	0.499	0.499	0.499	0.499	0.499	0.499	0.499	0.499	0.499
3.2	0.499	0.499	0.499	0.499	0.499	0.499	0.499	0.499	0.499	0.499
3.3	0.500	0.500	0.500	0.500	0.500	0.500	0.500	0.500	0.500	0.500
3.4	0.500	0.500	0.500	0.500	0.500	0.500	0.500	0.500	0.500	0.500

Table E2.1 Areas under the normal curve (to three decimal places)

231

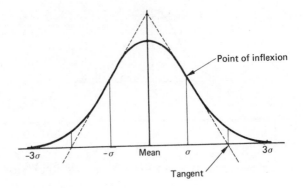

Fig. E2.9

iii) The places on the curve where the curvature changes (points of inflexion) occur where x has the values $\pm\sigma$.

iv) The slope of the curve at these points is such that the tangent to the curve where $x = \sigma$ should cross the x-axis where $x = 2\sigma$, and the corresponding tangent on the other side where $x = -2\sigma$.

The normal curve is completely symmetrical and it is convenient mathematically to take the value of the mean as the origin and to measure along the horizontal axis in intervals one standard deviation (σ). Areas below the curve between ordinates drawn at these intervals are given below and should be remembered:

> 68.26% of the total area lies below the curve between $\pm\sigma$
> 95.44% ,, ,, ,, ,, ,, ,, ,, ,, ,, $\pm2\sigma$
> 99.72% ,, ,, ,, ,, ,, ,, ,, ,, ,, $\pm3\sigma$

These areas are used in estimating *probabilities* with respect to frequency distributions which conform closely to the normal distribution.

Ogive
When a cumulative-frequency curve is drawn for a normal distribution, it gives a characteristic S-shaped curve known as an *ogive*.

Exercise E2
1   The table below shows the distribution in extensions produced in 130 springs by the same load. Draw a cumulative-frequency table and from it determine the percentage of springs in which the extension was less than 26.2 mm.

Extension (mm) (central value)	25.1	25.3	25.5	25.7	25.9	26.1	26.3	26.5
Frequency	2	4	11	28	39	24	15	7

2 A survey was conducted to find the number of rooms in each of 40 houses. Arrange the following results of the survey to form a cumulative-frequency table with one column for each of the house sizes occurring.

```
8  5  7  4  9  8  7  5
6  4  6  5  6  7  6  9
5  7  7  6  7  3  5  7
7  8  6  7  5  5  7  6
4  5  5  6  6  6  6  5
```

3 Seventy steel samples were analysed for nickel content. Draw the distribution curve for the values from the cumulative-frequency table below.

Nickel content (%)	2.2– 2.4	2.5– 2.7	2.8– 3.0	3.1– 3.3	3.4– 3.6	3.7– 3.9	4.0– 4.2	4.3– 4.5	4.6– 4.8
Cumulative frequency	3	5	10	27	48	61	65	68	70

4 Draw a distribution curve using percentage cumulative frequencies from data in the table below, which gives the results of tests on 150 cassette recorders.

Tape speed (cm/s)	4.775– 4.780	4.780– 4.785	4.785– 4.790	4.790– 4.795	4.795– 4.800	4.800– 4.805	4.805– 4.810	4.810– 4.815	4.815– 4.820
Frequency	3	7	19	28	47	22	16	6	2

5 Calculate the median weekly wage for the 80 people represented in the following table:

Weekly wage (£) (central value)	45	50	55	60	65	70	75
Frequency	3	6	11	33	21	5	1

6 Draw a cumulative-frequency curve using data from the table in question 5, and use the curve to estimate how many of the 80 people in the survey earn less than £64 per week.

7 The following table gives the distribution of crushing loads for 120 blocks of concrete. Draw a graph from data in this table using percentage cumulative frequencies. From the curve, determine the median crushing load.

Load (kN)	5.55– 6.05	6.05– 6.55	6.55– 7.05	7.05– 7.55	7.55– 8.05	8.05– 8.55	8.55– 9.05	9.05– 9.55
Frequency	5	11	28	42	19	11	3	1

8 Rainfall statistics for a period of 75 years are given in the table below. Draw a cumulative-frequency curve using this data and use it to determine the median and semi-interquartile range of the distribution.

Annual rainfall (cm)	36– 40	41– 45	46– 50	51– 55	56– 60	61– 65	66– 70	71– 75
Frequency	2	6	13	21	17	11	4	1

9 Draw a frequency distribution curve using the following data, which is taken from a survey of ticket sales on 300 aircraft. Does the curve have positive or negative skew? Would the value for the mean number of seats sold be greater or smaller than the median?

Tickets sold (central value)	25	75	125	175	225	275	325	375
Frequency	6	17	32	51	63	61	47	23

10 Which of the frequency distribution curves in fig. E2.10 represents the distribution with (a) the smallest range? (b) the largest mean value?

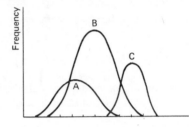

Fig. E2.10

11 What percentage of the total area below a normal distribution curve is enclosed between ordinates at (a) $+\sigma$ and -2σ? (b) $+3\sigma$ and $-\sigma$? (c) $-\sigma$ and -3σ?

E3 Probability

E3.1 What is probability?
In any problem or experiment, each separate result is called an *outcome*. The particular happening we are looking for will be called the *event*. Probability is the branch of mathematics which enables us to calculate the likelihood of any particular outcome.

We shall define the probability p of the event by

$$p = \frac{\text{number of ways the event can occur}}{\text{total number of possible outcomes}}$$

Two events are *complementary* when together they include every possible outcome without any overlap or duplication. When all possible outcomes are known, the set of outcomes can be referred to as the *sample space*.

Example When a couple have their first child, obviously it must be either male or female. Hence, for one child, the complete set of possible outcomes is {M, F}.

Assuming equal probability for each possible outcome, the probability of the first child being male is $\frac{1}{2}$ and the probability of it being female is equally $\frac{1}{2}$. Note that the probabilities of the complementary events add together to give 1.

Now consider a family with two children. The sample space is {MM, MF, FM, FF}. By symmetry, it is obvious that each of these possible outcomes, being equally likely, has a probability of $\frac{1}{4}$ (i.e. 1 in 4).

If we consider the possibility of such a family with two children having at least one boy, we can either add the individual probabilities for MM, MF, FM to get $\frac{1}{4} + \frac{1}{4} + \frac{1}{4} = \frac{3}{4}$, or we can use our basic definition from which

$$p = \frac{3 \text{ (ways the event can occur)}}{4 \text{ (possible outcomes)}}$$

If the probability that a certain event may happen is p, then the probability that the event will not happen is $1 - p$.

A probability of zero implies that there should be no chance at all of the event occurring. Note that negative probabilities are impossible and that p can never be greater than 1, since $p = 1$ implies a certainty. If we spin a coin with a head on both sides, for example,

$$p \text{ (tail)} = \frac{0}{2} = 0 \quad \text{and} \quad p \text{ (head)} = \frac{2}{2} = 1$$

The result will always be a head, never a tail.

E3.2 Expectation

Once we have determined the probability that an event will occur in any one trial, we can use it to predict how many times the event will occur in a number of similar trials. We define *expectation* as the product of the number of trials and the probability that an event will occur in any one trial.

Consider an experiment in which a pack of cards is shuffled and cut eight times. We can determine the expectation of clubs as follows:

$$p \text{ (club) in each trial} = \frac{13}{52} = 0.25$$

$\therefore$ Expectation in eight trials $= 0.25 \times 8 = 2$

This does not mean that there definitely will be two clubs when the cards are cut eight times, but that it is our best estimation based on knowledge of the chances that each trial will produce a club. (Note that the probability of the event occurring must be the same in each trial for our definition of expectation to apply.)

Example When 500 injection-moulded components taken at arbitrary intervals from a machine's production were given 100% inspection, 17 were found to be defective. Determine the probability that any single component will be defective and the expectation for defectives in a batch of 3000 components.

The probability that any given component will be defective is found using

$$p \text{ (defective)} = \frac{17}{500} = 0.034$$

In a batch of 3000 components, therefore,

$$\text{expectation of defectives} = 0.034 \times 3000$$
$$= 102 \text{ components}$$

E3.3 Dependent and independent events

Two events are independent when the outcome of one event has no effect on the probability of the outcome of the other. Spinning coins, throwing dice, and cutting a shuffled pack of cards are all examples of independent events. If the outcome of one event can affect the outcome of another, the two events are not independent.

Consider, for example, a bag containing three green beads and two blue beads. The probability of taking out a green bead at random is 3/5 and the probability of taking a blue bead is 2/5. If a bead is taken out and not replaced, the probability of a second bead being green may be 3/4 *or* 2/4 depending on whether the first bead taken was blue or green. These events are therefore not independent.

E3.4 Addition law for probabilities

When two events are *mutually exclusive*, the probability of one or the other occurring is calculated by *adding* their individual probabilities. We write

$$p(A \text{ or } B) = p(A) + p(B)$$

Mutually exclusive events

We say events are mutually exclusive if any one outcome prevents the possibility of another. For example, if a single card is taken from a pack, it may be a six or it could be a nine; but not both. Taking a six and taking a nine are mutually exclusive events. However, the single card could be a six and also a red card (e.g. the six of hearts), so taking a six and taking a red card are not mutually exclusive events.

In fig. E3.1, S represents the sample space and A and B are subsets of possible outcomes which do not overlap. The probability of an event being in S is 1; the probability of it being in A is $p(A)$; the probability of it being in B is $p(B)$. The event cannot be in both A and B and so the probability of the event being in A or B is given by

$$p(A \text{ or } B) = p(A) + p(B)$$

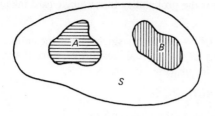

Fig. E3.1 Events which are mutually exclusive

Figure E3.2 shows the case of two events which are not mutually exclusive, when subsets A and B overlap. It is clearly possible for an event to be in A, or to be in B, or to be in both A *and* B. We can define these probabilities as $p(A)$, $p(B)$ and $p(AB)$ respectively. This duplication means that simple addition of probabilities will not be correct, since it implies that the area shared by A and B will be counted twice. The probability of an event being in either A or B or both becomes

$$p(A \text{ or } B) = p(A) + p(B) - p(AB)$$

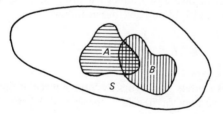

Fig. E3.2 Events which are not mutually exclusive

Example 1 What is the probability that a single card taken from a pack will be a six or a nine?

As there are four sixes and four nines in a pack,

$$p(A) = p(B) = \frac{4}{52} = \frac{1}{13}$$

The events are clearly mutually exclusive and so the probability that the card is a six or a nine is given by

$$p = p(A) + p(B) = \frac{2}{13} = 0.154$$

237

Example 2   What is the probability that a single card taken from a pack will be a six or a red card?

As in example 1, $p(A) = 1/13$ and, since half of the cards are red, $p(B) = \frac{1}{2}$. This time the events are not mutually exclusive, since the card taken can be both red and a six. In this case, therefore,

$$p = p(A) + p(B) - p(AB)$$

There are two red sixes in the pack, and so

$$p(AB) = \frac{2}{52} = \frac{1}{26}$$

$$\therefore \qquad p = \frac{1}{13} + \frac{1}{2} - \frac{1}{26} = 0.538$$

E3.5 Multiplication law for probabilities

The probability that two or more independent events which are not mutually exclusive will occur together is determined by *multiplying* their individual probabilities. We write

$$p(A \text{ and } B) = p(A) \times p(B)$$

For example, the probability of throwing a 3 with a die is 1/6. To determine the probability of throwing two threes one after the other, or at once with identical dice, we multiply probabilities as follows:

$$p(3 \text{ and } 3) = \frac{1}{6} \times \frac{1}{6} = \frac{1}{36} \text{ or } 0.028$$

Example   A man enters two raffles by buying three of the 100 tickets sold in the first raffle and one of 50 sold in the second. What is the probability that he will win both?

These events are not mutually exclusive, since the man may win both raffles. They are also independent, since a win in the first will not affect his chances of winning the second. The probabilities of winning the individual raffles are

$$p(1) = \frac{3}{100} \quad \text{and} \quad p(2) = \frac{1}{50}$$

Therefore his chance of winning both is given by

$$p(1 \text{ and } 2) = \frac{3}{100} \times \frac{1}{50} = \frac{3}{5000} \text{ or } 0.0006$$

E3.6 Practical probability problems

When solving probability problems involving more than one event, care must be taken in deciding whether to add or multiply probabilities. The multiplication law may be applied to events which are not completely independent provided we can determine how the outcome of one event will affect another. Sometimes we may have to use a combination of addition and multiplication laws to solve a problem. The following examples will illustrate some of these points.

Example 1   The probability that there will be no defectives in a batch of transistors is 0.73, and that for one defective per batch is 0.11. What is the probability that a batch will contain (a) not more than one defective? (b) more than one defective?

a) The probability of having no defectives $p(0) = 0.73$, and $p(1) = 0.11$. These are clearly mutually exclusive events, since there cannot be both 0 defectives and 1 defective in a batch. Using the addition law,

$$p(0) + p(1) = 0.73 + 0.11 = 0.84$$

This is the probability that a batch will contain not more than one defective.

b) The probability that there will be more than one defective in a batch is given by the addition law:

$$p(\text{more than } 1) = p(2) + p(3) + p(4) + \ldots + p(N)$$

where N is the batch size. Even if we knew these probabilities, it would be tedious to add them up for a large batch. We can instead use the fact that the probabilities of complementary events add together to give 1, so that

$$p(0) + p(1) + p(2) + \ldots + p(N) = 1$$

We have seen in part (a) of the question that $p(0) + p(1) = 0.84$, and so the sum of all the remaining probabilities is

$$p(2) + p(3) + \ldots + p(N) = 1 - 0.84 = 0.16$$

i.e. the probability that there will be more than one defective per batch is 0.16.

Example 2   Two dice are thrown together. What is the probability that the total score is 4?

There are three ways of scoring a total of 4: these are 3 and 1, 1 and 3, or 2 and 2. Consider the first combination, with 3 from the first die and 1 from the second. These events are clearly independent, since throwing the 3 will not affect the chances of obtaining a 1 with the second die. The probability of obtaining this result is found using the multiplication law. Both $p(3)$ and $p(1)$ are $1/6$, and so

$$p(3 \text{ and } 1) = \frac{1}{6} \times \frac{1}{6} = \frac{1}{36}$$

Similarly,

$$p(1 \text{ and } 3) = p(2 \text{ and } 2) = \frac{1}{36}$$

Since each of the three combinations is mutually exclusive, we obtain the total probability by addition, as follows:

$$p \text{ (score of 4)} = p \text{ (3 and 1)} + p \text{ (1 and 3)} + p \text{ (2 and 2)}$$

$$= \frac{1}{36} + \frac{1}{36} + \frac{1}{36}$$

$$= \frac{3}{36} \text{ or } 0.083$$

Example 3   A bag contains three green beads and two blue beads. If two beads are taken at random, what is the probability that they will both be green if (a) the first bead taken is replaced before taking a second? (b) the first bead is not replaced?

a) When taking the first bead, the probability that it is green is 3/5. Since the bead is replaced, the two events are independent and the probability of choosing the second green bead is also 3/5. Using the multiplication law, therefore, the probability of choosing two green beads is

$$p \text{ (two green. beads)} = \frac{3}{5} \times \frac{3}{5} = \frac{9}{25} \text{ or } 0.36$$

b) As in part (a), when taking the first bead,

$$p \text{ (green)} = \frac{3}{5}$$

However, by not replacing this green bead we affect the probability of taking a second green bead. There are now only four beads in the bag of which two are green, and so

$$p \text{ (second green)} = \frac{2}{4}$$

To find the probability of taking two green beads in this way we may still multiply the individual probabilities. Although the events are not completely independent (since leaving the first green bead out of the bag affects the probability of a second), the second probability is identifiable and the first event has no further effect on the second. We say the events are *conditionally independent*, and so

$$p \text{ (two green beads)} = \frac{3}{5} \times \frac{2}{4} = \frac{3}{10} \text{ or } 0.3$$

† **Example 4**   A card is taken from a shuffled pack followed by a second without replacing the first. Calculate the probability that the first card is an ace and the second a red card.

These events are not independent, since the ace we draw first could be a red card and thus affect the probability that the second is red. If we consider all the possible outcomes in turn and define their probabilities, we can arrive at an overall probability.

i) The first card taken is a red ace and

$$p \text{ (red ace)} = \frac{2}{52}$$

The second card is one of the 25 remaining red cards,

$$\therefore \quad p \text{ (red card)} = \frac{25}{51}$$

From the multiplication law,

$$p \text{ (red ace and red card)} = \frac{2}{52} \times \frac{25}{51} = \frac{50}{2652}$$

ii) The ace taken first is black and

$$p \text{ (black ace)} = \frac{2}{52}$$

The second card is red,

$$\therefore \quad p \text{ (red card)} = \frac{26}{51}$$

$$\therefore \quad p \text{ (black ace and red card)} = \frac{2}{52} \times \frac{26}{51} = \frac{52}{2652}$$

The combinations of cards described in (i) and (ii) are the only successful outcomes possible. They are mutually exclusive and so the total probability of taking an ace followed by a red card can be found by addition:.

$$p \text{ (ace and red card)} = \frac{50 + 52}{2652} = 0.038$$

Exercise E3

1 Define probability and state the range of possible numerical values.

2 If Manchester United play Everton at football, which of the following pairs of events is mutually exclusive?

  a) Manchester win and Everton win.
  b) Everton score three goals and Manchester win.

3 If the probability that an integrated circuit will be damaged during soldering operations is 0.0012, what is the expectation of damaged components on 1000 printed circuit boards each containing 20 integrated circuits?

4 If the probability that a refrigerator fails because of a faulty thermostat is 0.57, what is the total probability of other causes of failure?

5 A box contains 100 discs numbered 1 to 100, from which two are taken at random. Which of the following events are completely independent?

a) Taking an even numbered disc followed by a disc between 1 and 50 if the first is replaced.
b) Taking an even numbered disc followed by an odd one if the first is replaced.
c) Taking an even numbered disc followed by a disc between 1 and 50 if the first is not replaced.
d) Taking an even numbered disc and then, having replaced it, taking a second disc with a smaller number.

6 State the addition law for probabilities when events are mutually exclusive and use it to determine the probability of throwing either a 5 or an even number with a single die.

7 State the addition law for probabilities when events are not mutually exclusive and use it to find the probability of throwing either a score of 8 or a throw including at least one 6 with a pair of dice.

8 The probability that a resistor taken from a batch will have a value greater than 508 Ω is 0.03, while the probability that it will be less than 446 Ω is 0.12. Determine the probability that the resistance will lie between these two values.

9 State the multiplication law for probabilities and hence find the probability of throwing two sixes with a pair of dice.

10 A man travels by bus to a railway station with a probability of 0.85 that he will arrive in time for his train. If the probability that his train is cancelled is 0.06, what is the probability that he will arrive at his final destination on time, assuming no further delays?

11 Five discs numbered 1 to 5 are placed in a bag and two are taken at random without replacement. What is the probability that the discs have numbers which differ by one?

12 In a box of ten capacitors, three are defective. One capacitor is taken out at random, replaced, and then another is taken. What is the probability that both are defective?

13 What would be the probability of taking two defective capacitors in question 10 if the first was not replaced before taking the second?

14 A box contains five steel nuts, four brass nuts, three steel washers, and two brass washers. If one item is taken at random and then a second without replacement, what is the probability that the first item was a nut and the second made of brass?

Revision exercises

Revision exercise A1
Use a calculator to evaluate the following expressions.
1   $(3.181 + 2.753) \times 1.877$
2   $\dfrac{16.23}{1.77} - \dfrac{9.58}{0.96}$
3   $(3.23)^2 + \sqrt{73.25}$
4   $e^{-2.15}$
5   $\ln(8.76/3.92)$
6   $\log(\sqrt{82.4})$
7   $\sin 32.13°$
8   $\tan 66° \, 42'$
9   $\arccos 0.7235$ (one value between $0°$ and $90°$)
10   $\sec x$, where $x = 1.165$ rad
11   $1/2\pi \sqrt{(LC)}$, where $L = 0.520$ and $C = 0.006$
12   11.5% of 118.6
13   $e^{-\log 8.66}$
14   $28.75^{1.25}$
15   $\sqrt[1.8]{216.6}$
16   Draw a flow chart to show the calculator steps you would use to determine values of V where $V = e^{-ct}$ for $t = 0$ to $t = 4.5$ in steps of 0.5. (c is a constant.)

Revision exercise A2
1   Use a calculator or tables to solve the following equations to find x:
(a) $1.61 \log x = 1.22$ (b) $2.67 \ln x = 9.25$

2   The voltage across an inductor is given by $v = Ve^{-Rt/L}$. Use a calculator or tables to find R if $v = 53.14$ when $t = 0.015$, $V = 65$, and $L = 0.35$.

3   The tension on one side of a belt drive is given by $F_1 = F_2 e^{\mu\theta}$. Calculate μ if $F_1 = 100$, $F_2 = 60$, and $\theta = 3.77$.

Revision exercise B1
1   From the formula $t^2 = (8L + 3Mg)/8bdY$,
a) find the value of t when $L = 3.5$, $M = 12$, $g = 9.81$, $b = 0.78$, $d = 0.022$, and $Y = 1.35 \times 10^7$;
b) find a formula for L in terms of the other quantities.

2   From the filtration formula $Pt = mv^2/A^2 + bV/A$,
a) find the time (t) for filtration when $P = 38$, $V = 3.2$, $A = 720$, $m = 175\,000$, and $b = 1800$;

b) find a formula for m in terms of the other quantities.

3   If frictional loss is given by $F = 2fLV^2/Dg$,

a)   find F when $f = 0.0054$, $L = 60$, $V = 25$, $D = 0.04$, and $g = 9.81$;

b)   deduce a formula for V in terms of the other quantities.

4   From van der Waals' equation $(P + a/V^2)(V - b) = nRT$, find P when $a = 4.17$, $V = 1.25$, $b = 0.0371$, $n = 2.00$, $R = 0.0848$, and $T = 298$.

Revision exercise B2

1   If V is the potential drop over a length l along a straight uniform wire such that $V \propto l$, find the potential drop over 800 mm given that V is 1.2 volts when $l = 240$ mm.

2   The safe uniformly distributed load (L) on a particular type of steel girder is inversely proportional to the span (S). If the safe load is 1920 kg on a span of 7.5 m, find the possible increase in L if S is reduced by 3 m.

3   According to Charles' law, if a fixed mass of gas is kept at constant pressure its volume is proportional to its temperature measured on the Kelvin scale. Find the increase in volume if 25 litres of gas at constant pressure and a temperature of 300 K are heated to 375 K.

Revision exercise B3

1   Under certain conditions, the bending moment M of a loaded beam is given by $M = 0.25(3.2 + x)^2 - 4.6x$, where x is the distance from one of the supports. Calculate the values of x for which $M = 0$.

2   From the formula $2Am(n - c) + Bn^2 = 0$, find n if $A = 1.65$, $B = 6.5$, $C = 8.4$, and $m = 14.5$.

3   The shape of an arch is an arc of a circle of radius R. If the span is S and the rise h, prove that $h^2 - 2Rh + 0.25S^2 = 0$ and hence find h when $S = 3.6$ m and $R = 2.2$ m.

4   For a third-order chemical reaction, the equation $k = \dfrac{x(2a - x)}{2a^2 t(a - x)^2}$

gives the reaction-rate constant k in terms of the time t it has taken for a quantity x of a chemical to be used up from an initial concentration a. Find x when $k = 0.096$, $a = 0.25$, and $t = 150$.

Revision exercise B4

1   Simplify each of the following expressions as far as possible:

(a) $(2x^2)^3$ (b) $12a^4 \div 3a^3$ (c) $\sqrt{16b^4}$ (d) $3c^{2+x} \div 2c^x$
(e) $(64d^6)^{1/3}$

2   Solve the following indicial equations without using logarithms:

(a) $4^{2x+1} = 1024$ (b) $49^x = 343^{x-1}$ (c) $64^{3x}/16^{x-1} = 4$

3   Find both solutions to each of the following equations, without using logarithms:

(a) $3^{x^2-3} = 9$ (b) $12^{3x} = 144^{x^2-1}$ (c) $9^{t+1} = 81^{1/2t}$

4   Solve the following pair of simultaneous indicial equations without using logarithms: $4^{x+y} = 64^{x-2}$ and $3^{x-y} = 9^{x-3}$.

Revision exercise B5

1 A vehicle begins to accelerate from an initial speed u and the following speeds were recorded at 3s intervals:

Time t(s)	3	6	9	12	15
Speed v(m/s)	8.1	15.0	21.9	28.8	35.7

Plot a graph of v against t to show that $v = u + at$, where a is the acceleration and a is a constant. Use your graph to determine u and a.

2 A copper bar is 1.850 m long at 0°C. Its length is measured as it is heated, giving the following results:

Length l(m)	1.850	1.852	1.855	1.859	1.869
Temperature θ(°C)	0	50	150	300	600

Draw a graph of temperature against length and show that $l = L + \alpha\theta L$, where L is the original length and α(/°C) is the coefficient of linear expansion. Find α from your graph.

3 A straight-line graph has the equation $s = a + bt$. If $s = 1.6$ when $t = 32.9$, and $s = 13.8$ when $t = 145.3$, calculate the values of a and b. What is the value of t when $s = 2.9$?

4 An electric motor is connected to a d.c. supply. The back e.m.f. E is given by the equation $E = V - IR$, where I is the armature current, R is the armature resistance, and $V = 12$ V. Use the following table of values to plot a graph of E against I and use the graph to determine R in ohms.

E (volts)	3.2	4.7	6.2	8.3	11.1
I (amperes)	10.4	8.6	6.9	4.4	1.1

Revision exercise B6

1 Using values from the following table, plot a graph of y against $1/x^2$ to show that $y = M/x^2 + C$. Determine values of M and C from your graph.

x	2.00	3.00	4.00	5.00	6.00
y	8.33	6.09	5.31	4.94	4.75

2 Using values from the following table, plot the best straight line to show that $s = K\sqrt{t}$. Determine the value of K from your graph.

t	3.61	4.82	5.55	6.82	7.30
s	5.41	6.24	6.70	7.41	7.68

3 Use the following table to draw a suitable straight-line graph showing that $y = ax^n$. Find values of a and n from your graph.

x	2.0	4.0	6.0	8.0	10.0
y	21.25	82.11	181.03	317.24	490.19

245

Revision exercise B7

1 Solve graphically the simultaneous equations

$$y = 8.8x - 1.6 \quad \text{and} \quad y = 4.1x + 3.3$$

2 The following table gives values of y_1 and y_2 for a range of values of x. If $y_1 = ax + b$ and $y_2 = cx + d$, show graphically that the lines intersect at $(1.2, 6.8)$ and determine the values of $a, b, c,$ and d.

x	0.2	0.4	0.6	0.8	1.0
y_1	7.8	7.6	7.4	7.2	7.0
y_2	5.3	5.6	5.9	6.2	6.5

3 Solve the following simultaneous equations using the method of substitution: $y = 1.5x - 3.2$ and $3.5y = 7.2x + 1.9$.
4 Solve the simultaneous equations $1.2y = 4.4x - 1.8$ and $5.7y = 6.6x + 9.6$ by eliminating x.
5 Plot the graph of $y = 1.5x^2 + 2.6x - 3.1$ for values of x between -4 and $+2$. Use your graph to solve the equations

a) $1.5x^2 + 2.6x - 3.1 = 0$
b) $1.5x^2 + 2.6x - 1.1 = 0$

6 Use a graph to find the roots of the equation $x^3 - 2.6x^2 - 3.6x + 6.3 = 0$.

Revision exercise B8

1 Plot a straight-line graph from the values in the following table, using log–linear graph paper. Show that $y = ae^{bx}$ and estimate values for the constants a and b.

x	1.00	2.00	3.00	4.00	5.00
y	3.66	6.10	10.16	16.92	28.18

2 The torque T required when drilling aluminium with a fixed feed rate is given by $T = ad^n$, where d is the drill diameter and both a and n are constants. Plot the following experimental results on log–log graph paper to give a straight line. Determine a and n from your graph.

d (mm)	4.00	6.00	9.00	15.00	30.00
T (Nm)	0.04	0.09	0.18	0.46	1.60

Revision exercise B9

1 Convert to polar co-ordinates (a) $(6.5, 2.9)$; (b) $(-2.3, 11.4)$; (c) $(3.7, -1.66)$; (d) $(-28.5, -19.7)$.
2 Convert to Cartesian co-ordinates (a) $(12.5, 36.9°)$; (b) $(7.4, 62° 26')$; (c) $(18.8, \pi/3)$; (d) $(32.2, -48.5°)$.
3 Plot the points $(2.8, 55°)$, $(3.6, 138°)$, $(6.2, 215°)$, and $(4.3, -67°)$ on the same polar graph.
4 Draw the spiral $r = 2\theta$ between θ and 0 and $\theta = 2\pi$ rad, using points drawn every $\pi/6$ rad. Measure r in mm and draw your graph five times full size.

5 The range and angle of elevation of a satellite returning to earth are measured by radar from the time it passes directly overhead until its altitude is too low to be measured. Use the following table to draw a polar graph showing the path of the satellite, and estimate how far from the radar station it will land when $\theta = 0°$.

Elevation θ (degrees)	90.0	75.0	60.0	45.0	30.0	15.0
Range r (km)	63.8	56.5	53.7	54.0	57.2	62.5

Revision exercise B10
1 If any switch in fig. RQ1 is considered to be closed when its input is 1, determine whether the lamp will be on or off for the following inputs:

a) $A = 1, B = 0, C = 1, D = 1$
b) $A = 0, B = 1, C = 0, D = 1$
c) $A = 1, B = 1, C = 1, D = 0$

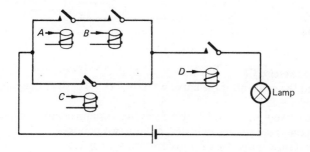

Fig. RQ1

2 Write down the output $T = 1$ or $T = 0$ for the circuits in fig. RQ2 with the inputs given.

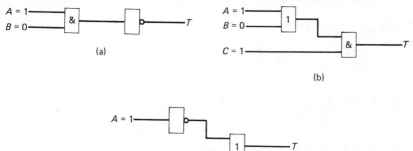

Fig. RQ2

3 Complete the truth table in fig. RQ3 for all possible combinations of A and B.

A	B	$A+B$	$\overline{A}.\overline{B}$	$A+\overline{B}$
0	0			
0	1			
1	0			
1	1			

Fig. RQ3

4   Use a truth table to demonstrate that $\overline{A.B} = \overline{A} + \overline{B}$. (Note that the truth-table values for $\overline{A.B}$ are found by inverting values for $A.B$).

5   Draw a truth table to show that $A + B.C = (A + B).(A + C)$.

6   Figure RQ4 shows a circuit made from three logic gates. Draw a truth table to show the output T for all possible values of A and B. What type of gate with inputs A and B and output T would replace this circuit completely?

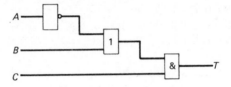

Fig. RQ4

Revision exercise C1

1   Find the gradient of the curve $y = x^3 - 2x^2 + 3x - 4$ at the points where $x = 1$ and $x = 2$.

2   The curve $y = x^3 - 6x^2 + 9x + 5$ has two turning points: a minimum and a maximum. Find the co-ordinates of these two points.

3   If $y = \sin x$, prove that $y + dy/dx = \sin x + \cos x$.

4   If $y = 4x^3$, show that $x(dy/dx) - 3y = 0$.

5   Given $y = x^3 - 3x^2 + 6x - 6$, prove that $y + dy/dx = x^3$.

6   Prove that the equation $y + x(dy/dx) = 0$ is satisfied by $y = 1/x$.

Revision exercise C2

1   Find (a) $\int 4\sqrt{x}\ dx$, (b) $\int v^{1.4}\ dv$, (c) $\int z^{-2.3}\ dz$.

2   Evaluate (a) $\int_4^9 x^{3/2}\ dx$, (b) $\int_0^{\pi/3} (1 + \cos\theta)\ d\theta$.

3   Find the value of $\int_1^9 (3 - x)^2\ dx$.

4   If $\int_0^x (2t + 1)\ dt = 6$, where x is positive, find the value of x.

5   Given that $\int_0^x 4\cos\theta\ d\theta = 2$, where $0 < x < \pi/2$, find the value of x.

Revision exercise D1

1   If a cylinder has a height equal to its diameter and a volume of 2155 mm³, find its diameter.

2   Find the percentage of metal saved if two pipes with internal diameter 25 mm and wall thickness 2.5 mm are replaced by a single pipe with an equivalent internal cross-sectional area and wall thickness 3.0 mm.

3   A feed hopper is in the form of a frustum of a pyramid with an open square top of side 1.8 m and a square base of side 0.6 m. Find the capacity of the hopper if its vertical height is 2.2 m.

4   A cylindrical tank lies with its axis horizontal. If its diameter is 1.6 m and its length 2.8 m, calculate the volume of liquid which fills the tank to a depth of 1.2 m.

5   A concrete pillar is made of two parts. The top section is a cylinder of radius 120 mm and height 800 mm. The base is a frustum of a cone with base radius 250 mm, top radius 120 mm, and height 200 mm. Find the total volume of the pillar.

6   The ball of a float valve is a sphere of diameter 200 mm. Find the area of the ball in contact with water if the ball is immersed to a depth of 30 mm.

7   Calculate the area wasted when 120 squares of side 5 mm are cut from a 70 mm diameter silicon disc during the manufacture of integrated circuits.

8   Calculate the volume of gas above the piston in fig. RQ5 if the top of the cylinder is spherical.

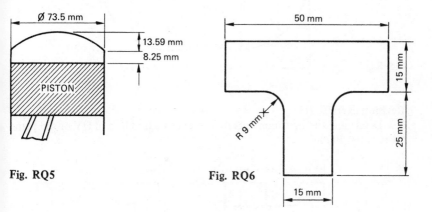

Fig. RQ5   Fig. RQ6

9   Determine the volume of a 5.5 m length of the T-section bar shown in fig. RQ6.

Revision exercise D2

1   Given that $y = \cos^2 x$, draw up a table of values of x and y, taking x from 0° to 90° at intervals of 15°. Use these values of y and the strip width of $\pi/12$ radians with Simpson's rule to show that the area under the curve over this range is $\pi/4$. Sketch the curve.

2 For the graph of $x^2 + 4y^2 = 100$, calculate the values of y which correspond to the following values of x:

$$-10, -8, -6, -4, -2, 0, 2, 4, 6, 8, 10$$

Use Simpson's rule to find the area of this ellipse.

3 For a uniformly loaded cantilever of length 8 m it is known that for a particular load the deflection in millimetres at the free end is given by the area under the graph of

$$y = \frac{x^3 - 24x^2 + 192x}{120}$$

for values of x from 0 to 8. Use Simpson's rule to calculate the deflection.

4 Figure RQ7 shows the outline of a rotary engine chamber. Use Simpson's rule to determine the cross-sectional area of the chamber from the approximate dimensions given.

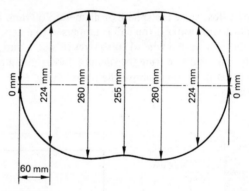

Fig. RQ7

Revision exercise D3

1 Determine the first moments of area about axes OX and OY of the plane areas in fig. RQ8.

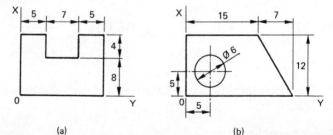

Fig. RQ8   (a)   (b)

2 Find the height of the centroid above the base of the cross-sectional area in fig. RQ9.

3 By how much will the centroid of the lamina in fig. RQ10 be moved, and in which direction, if a strip 8 mm deep is cut from the top as shown?

250

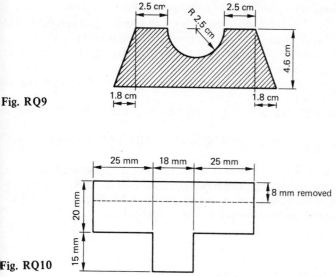

Fig. RQ9

Fig. RQ10

Revision exercise D4

1 A support column is 600 mm high. The base and top are 250 mm in diameter and the column curves uniformly from each end to a central diameter of 150 mm. Use the prismoidal rule to find the volume of this column.

2 An excavation for an arena in level ground is elliptical in plan and 20 metres in depth, with uniformly sloping sides. If the level bottom is an ellipse 200 m x 120 m and the top is an ellipse 230 m x 150 m, find the length of the top perimeter and use the prismoidal rule to find the volume of the excavation.

3 A cylindrical bar of diameter 25 mm has a groove cut round its circumference. If the cross-section of the groove is a semicircle of radius 3 mm, find the volume of material removed.

4 The cross-section of an underground railway tunnel is the major segment of a circle, the base chord being 3.2 m and the height at centre also 3.2 m. The tunnel follows an arc of mean radius 200 m, and the arc subtends an angle of 12° at the centre of curvature. Calculate the volume of air in this section of the tunnel.

Revision exercise D5

1 If ABC is a triangle right-angled at C, show that $\sin A + \cos A + \tan A + \cot A = \sin B + \cos B + \tan B + \cot B$.
2 Prove that $\sin A \cos A \sec A \csc A = 1$.
3 Show that $\sin x(1 - \cot x) + \cos x(1 - \tan x) = 0$.
4 Evaluate $3 \sin 106° + 5 \tan 72° + 4 \cos 218°$.
5 Evaluate $3 \sin x + 4 \cos x - 6 \tan x$ when $x = 203°$.
6 Find the value of $\sin x(1 - \cos 2x) + \cos x(1 - \sin 2x)$ when $x = 0.34\pi$.

Revision exercise D6

1 Silicone-injection holes are drilled into a wall which is 280 mm thick. The drill is inclined at an angle of 35° to the horizontal and is held in a vertical plane which is perpendicular to the wall surface. To what depth must the holes be drilled to reach the centre of the wall?

2 Find the area of triangle ABC in which AB = 11.5 mm, BC = 16.9 mm, and the angle at C is 54.6°.

3 In triangle XYZ, side XY = 32 m, YZ = 66 m, and angle $X = 28°$. Find angle Z.

4 Calculate the area of triangle ABC in which AB = 18.2 mm, angle $A = 65°$ and angle $C = 33°$.

5 To reach a town Q from town P, which lies on the opposite side of a lake, a car travels 3 km due E along a straight road from P to R. At R, the car joins a second straight road and travels a further 2 km in a NW direction from R to Q. Calculate the distance between P and Q across the lake.

Revision exercise D7

1 Sketch the shapes of the following curves for values of θ from 0 to 2π radians:

(a) $y = 1 + \sin \theta$ (b) $y = 2 - \cos \theta$ (c) $y = 1 - \sin^2\theta$

2 Draw up a table of values of x and y over the range of $0 \leqslant x \leqslant 360°$ for the equation $y = 2 \sin (x + 30°)$ and plot the curve. Mark on the graph the distances representing amplitude and phase angle.

3 Plot the graph of $y = 3 \sin x + 4 \cos x$ over one cycle. Hence find the maximum value of $3 \sin x + 4 \cos x$ and the value of x for which $y = 0$.

Revision exercise D8

1 Use standard trigonometrical identities to show that

$$\frac{1}{(1 + \cot^2 x)(1 + \tan^2 x)} = \sin^2 x \cos^2 x$$

2 Show that $\dfrac{\csc^2 x \tan x}{\sec^2 x} = \cot x$.

3 Find all the values of x between $x = 0°$ and $x = 360°$ which are solutions of the equation $2 \sin x = 0.46$.

4 Solve the equation $\sec \theta = 2.495$, giving values of θ in radians within the range $-2\pi \leqslant \theta \leqslant 2\pi$.

5 Find all solutions of the equation $3 \cos^2 \theta = 2.281$ between $\theta = 0°$ and $\theta = 180°$.

6 Determine the four values of x in the range $-180° \leqslant x \leqslant 180°$ which satisfy the equation $3.2 \sin^2 x + 1.1 \sin x - 1.6 = 0$.

Revision exercise D9

1   An aeroplane is heading due north with a velocity of 480 km/h when it encounters a cross wind blowing due east (090°) with a velocity of 88 km/h. Find the actual speed and direction of the plane.

2   Find the resultant of the following forces, all acting at the same point A and taken in order clockwise from AB: AB = 7.2 N vertically upward; AC = 5.4 N and $\angle BAC = 60°$; AD = 2.8 N and $\angle CAD = 45°$; AE = 3.6 N and $\angle DAE = 90°$.

3   A ship which is heading due south (180°) at 20 knots is subject to a cross-current moving due east (090°) such that its resultant bearing is 167°20'. Find the speed of the cross-current.

Revision exercise E1

1   A batch of 100 solid stone slabs for paving were weighed (correct to the nearest kilogram) and the following results were obtained:

Mass in kg	6	7	8	9	10	11	12	13	14
No. of slabs	1	2	5	28	39	20	3	1	1

Construct a histogram to illustrate these figures. Calculate the arithmetic mean and the standard deviation.

2   The time by various students to complete a titration experiment varies as shown in the table below. Calculate the mean and the standard deviation for these results.

Time (seconds ±10)	210	230	250	270	290	310
No. of students	2	22	40	18	7	1

3   In a survey of 200 evening classes, the following figures were obtained for class size. Calculate the mean and the standard deviation.

Enrolment	10	11	12	13	14	15	16	17	18	19
No. of classes	1	3	5	17	31	53	49	26	11	4

Revision exercise E2

1   Draw a cumulative-frequency curve using values from the following table of weekly egg-production figures. Find the median egg mass from your curve.

Mass (g)	40.0–44.9	45.0–49.9	50.0–54.9	55.0–59.9	60.0–64.9	65.0–69.9	70.0–74.9
Number of eggs in one week	13	46	75	89	72	51	28

2   The lengths of 400 nylon bushes machined on an automatic lathe were measured to the nearest 0.1 mm. Use the following table of results to draw a cumulative-frequency curve and determine the median length.

Length (mm) (class boundaries)	11.95–12.15	12.15–12.35	12.35–12.55	12.55–12.75	12.75–12.95	12.95–13.15	13.15–13.35	13.35–13.55
Frequency	9	26	73	157	69	36	19	11

3 Calculate the median number of defects in used cars from the following table of values resulting from tests on 300 cars.

Number of defects (central values)	1	4	7	10	13	16
Frequency	42	42	52	83	53	28

4 Draw a frequency curve using values from the following table obtained during tests on 280 zener diodes. From the shape of the curve, would you expect the mean voltage to be higher than the median or lower?

Voltage (V) (central values)	8.45	8.50	8.55	8.60	8.65	8.70	8.75	8.80
Frequency	14	44	54	51	43	34	24	16

5 Sketch a normal curve and draw two vertical lines at equal distances, one on each side of the mean, so that the area enclosed below the curve and between the lines is one half of the total area below the curve. How many standard deviations are these lines away from the mean?

Revision exercise E3

1 Which of the following pairs of outcomes is mutually exclusive?

a) Throwing a three and throwing an even number with one throw of a die.
b) Taking a king and taking a black card, when a single card is taken from a pack.

2 The probability that any individual onion seed fails to germinate is 0.04. What is the expectation of failure if 350 seeds are sown?

3 Two runners, A and B, enter a race. The probability that A will win is 0.14 and that B will win is 0.18. What is the probability that either A or B will win?

4 The probability that a carton of electrical components contains no defectives is 0.79 and that of one defective is 0.12. Determine the probability that there will be (a) less than two defectives, (b) one or more defectives.

5 The probability that the next bus to arrive at a bus stop will be a number twelve is 0.29. The probability that any bus will arrive full is 0.14. What is the probability that the next bus to arrive will be a number twelve which is not full?

6 The contents of eight identical, unmarked cans of paint are as follows:

3 white undercoat
2 white gloss
2 red undercoat
1 red gloss

If two cans are opened, what is the probability that the first is red paint and the second is undercoat?

Answers to numerical exercises

Exercise A1

1	11 493	16	7.879
2	22.606	17	15.96
3	6542	18	1.30
4	5.876	19	1.22
5	133 293	20	0.852
6	1.7504	21	0.843
7	6.415	22	3.73
8	43.48	23	32
9	0.4649	24	57.91
10	27.42	25	4.08
11	£368	26	0.365
12	$7.50	27	1.06×10^4
13	2.814	28	(a) 18.62, (b) 165.78, (c) −0.59,
14	3.624		(d) 0.555
15	0.907		

Exercise A2

1 (a) 2.0794, (b) 3.4657,
   (c) 1.0717, (d) −4.8536
2 0.66
3 (a) 2.458, (b) 32.14

4 1.22
5 50.1 μF
6 58.91 h

Exercise B1

1 (a) $x = z - 4y$,
   (b) $s = \frac{1}{3}(R + 4T - Q)$,
   (c) $c = 2(a - 5d - b)$

   (d) $y = z(a - b) - x$,

   (e) $F = \frac{9}{5}C + 32$,

2 (a) $R = \frac{V}{I}$, (b) $p_2 = \frac{p_1 V_1}{V_2}$,

   (c) $y = \frac{fI}{M}$, (d) $r = \frac{100I}{Pn}$,

   (e) $x = \frac{WL}{AE}$, (f) $t = \frac{v - u}{a}$

   (f) $d = e\left(\frac{a - b}{c} + 2fg\right)$,

   (g) $u = 4\left[t - \frac{p(r + s)}{q}\right] - v$

3 (a) $z = \frac{1}{2}[5(x - w) - 3y]$,

   (b) $F = \frac{2m}{t^2}(s - ut)$

   (c) $R_1 = \frac{RR_2}{R_2 - R}$,

4 (a) $r = \sqrt{\frac{A}{\pi}}$, (b) $L = \frac{gT^2}{4\pi^2}$,

   (c) $d = \sqrt[3]{\frac{12M}{b}}$,

   (d) $c = \frac{b^2 - d^2}{4a}$,

(e) $z = \sqrt{(x^2 - y)} - k,$

(f) $L = M(J + 1)^{\frac{2}{3}} - K$

5 (a) $b = \dfrac{a}{1 - e}$, (b) $x = \dfrac{z - w}{y - 2w}$,

(c) $S_2 = \dfrac{K_1 S_1}{K_1 + K_2}$,

(d) $x = \dfrac{cd}{b - a - c}$,

(e) $h = \sqrt[3]{\dfrac{12M}{b}}$,

(f) $n = \dfrac{\cdot C}{\sqrt{(\sqrt{P} - 1)}}$

6 470 Ω
7 7.66 m/s
8 6.30 mm
9 11.5 Ω

Exercise B2

1 (a) $L = \text{f}(\theta)$; θ is independent variable
   (b) $t = \text{f}(D)$; D is independent
3 (a) 2.5, (b) 27, (c) 0.85

4 6 N
5 $F \propto 1/D^2$, $F = k/D^2$
6 4.47 Hz, 223.6
8 0.054, 6760 mm^4

Exercise B3

1 a) $(x + 2)(x + 3)$
   b) $(m - 4)(m + 5)$
   c) $(x - 2)(x - 6)$
   d) $(2z + 1)(z + 3)$
   e) $(3x + 2)(5 - x)$
   f) $(2r + 1)(3r - 4)$
   g) $(1 - 3x)(4x + 5)$
2 a) 4 and 3
   b) 6 and $-3/2$
   c) -2 and $3/5$
   d) $2/3$ and -5
   e) $-1/5$ and 2
   f) 3 (equal roots)
   g) 6 and $-5/2$
   h) -1 and -3

3 a) 0.85 and -2.35
   b) 1.11 and -2.74
   c) 8.5 (equal roots)
   d) -8.63 and 4.63
4 4
5 1.61
6 1.01 s
7 17 Ω
8 5.41 mm and 8.41 mm
9 2 m
10 60 mm
11 119.6 m
12 0.016

Exercise B4

1 (a) 7^3; (b) 8^3, 2^9; (c) 9^4, 3^8; (d) 20^2
2 (a) x^5, (b) $x^{5.5}$, (c) a^2, (d) x^{3y-2}, (e) b, (f) $y^{1.5}$, (g) z^{-3} or $1/z^3$
3 (a) x^{12}, (b) $v^{2.76}$, (c) a^{-bc} or $1/a^{bc}$, (d) $x^{\frac{1}{5}}$, (e) $y^{1/x}$, (f) $s^{2/t}$, (g) b

4 (a) 3^{3x}, (b) 4^{6x}, 8^{4x}, (c) 5^y, (d) $3^{4/x}$
5 (a) 2, (b) $\frac{3}{4}$, (c) $\frac{4}{5}$, (d) $\frac{4}{3}$
6 (a) ± 1, (b) ± 0.707, (c) 4 or -2, (d) 3 or -1, (e) 1.57 or -0.32
7 (a) $x = 0.833$, $y = -0.50$
   (b) $x = 0.421$, $y = 0.526$

Exercise B5

2 18 years
3 0.20 kg/K, −40 kg
4 4.4 volts, 15 ohms
5 (a) 2, 1.5; (b) −5, 7; (c) 1.5, 2.25; (d) 0.25, −0.625; (e) 1.8, 0.4; (f) −1, 7; (g) 2, 13; (h) 0.8, 4

6 6.23 sq. units
8 5°C, 1°C/s
9 0.5, 3, 4
10 250 Pa, −0.2 Pa/K

Exercise B6

1 3.6, 0.2
2 945
3 0.634

4 2.5, −8.5
5 1, 4
6 2.4, 0.5

Exercise B7

1 $x = 2.5, y = 11.8$
2 $t = 7.5, v = 58.25$
3 (a) $x = 2.38, y = 5.75$
 (b) $v = 0.92, t = -2.25$
5 3.75, 1.6; 3.07, 2.28; 4.71, 0.64
6 4.45, −0.45; 3.73, 0.27; −4
7 2.82, 0.18; 3.41, 0.59

8 (a) 1.62, −0.62; (b) 0.73, −2.73
9 $x = 0.18, y = 2.64$; $x = 1.82, y = -0.64$
10 (a) 1.17, 6.83, −3.50
 (b) 3.20, 5.19, −0.19
 (c) 1.20, 6.40, −52.0
11 3.25

Exercise B8

2 0.72, 1.20
3 760, −6.5 x 10⁻⁵
4 0.116 seconds

5 23, 1.85
6 1.41, 52.0

Exercise B9

1 (a) (7.74, 57.13°);
 (b) (16.12, 119.74°);
 (c) (0.37, 251.08°) or (0.37, −108.92°);
 (d) (276.83, 296.84°) or (276.83, −63.16°)

2 (a) (1.87, 8.8); (b) (−1.21, 0.31);
 (c) (−28.15, −16.25);
 (d) (0.69, −0.35);
 (e) (5.38, −14.00)
3 6.28 km west, 3.03 km north
4 −36.26 mm x, 99.63 mm y
8 (49.5, 194.5°)

Exercise B10

1 (a) off, (b) off, (c) on, (d) on
2

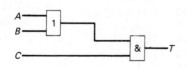

3 (a) 0, (b) 1, (c) 0
4 (a) 0, (b) 1, (c) 0

6

A	B	T
0	0	1
0	1	1
1	0	0
1	1	1

(a)

A	B	C	T
0	0	0	0
0	0	1	0
0	1	0	1
0	1	1	0
1	0	0	0
1	0	1	0
1	1	0	0
1	1	1	0

(b)

(c)

A	B	T
0	0	1
0	1	0
1	0	0
1	1	1

8 (a)

$A + \bar{B}$
1
0
1
1

(b)

$\bar{A}.B$
0
1
0
0

7

A	B	C	$(A + \bar{B}).\bar{C}$
0	0	0	1
0	0	1	0
0	1	0	0
0	1	1	0
1	0	0	1
1	0	1	0
1	1	0	1
1	1	1	0

(c)

$A + \bar{B} + C$
1
1
0
1
1
1
1
1

Exercise C1

1 3, 12, 27, 13
2 $3x^2$
3 $4x$
4 (a) $6x^2$, (b) $10x$, (c) $2x + 6$, (d) $6x - 12$
5 $36 + 9.8t$
6 10
7 0.5
8 $3 - 2x + \cos x$
9 (a) $15x^2$, (b) $7 - 6t$, (c) $1 + \cos \theta$
10 (a) $6(\sqrt{x} + 1/x^3)$, (b) $1.8\, Pr^{0.8}$
11 3, 3.5
12 2

Exercise C2

(Note: Arbitrary constants have been omitted)

1 (a) $2x^2$, (b) $2x^4$, (c) $0.5x^2 + 2x$, (d) $2x^3 - 2x^2 + 3x$
2 $3x^3 + 3x^2 - 10x + 4$
3 122.5
4 (a) $3x^3 - x^4$, (b) $2x^2 - 8x$, (c) $0.6x^3 - 1.1x^2$
5 (a) $4\pi r^3/3$, (b) $4t - 3.6t^3$
6 (a) 1.0, (b) 56.0
7 2.0
8 20.83 sq. units
9 $y = x^3 - 4x^2 + 5x + 2$

Exercise D1

1 5122 mm^2
2 484.6 mm^2
3 5.43×10^3 mm^2, 2.90×10^4 mm^3
4 4.50 mm^3
5 25 g
6 0.508 m^3, 4.226 m^2
7 4565
8 1.96×10^4 mm^3
9 (a) 452 mm^2, (b) 16.43 m^2, (c) 175.9 mm^2
10 (a) 3.41×10^4 mm^3, (b) 1.57×10^4 mm^3
11 374 mm
12 4.91×10^4 mm^3
13 27.96 kg
14 513 m^3/h
15 1.88 km^2
16 69.1 g

Exercise D2

1 70 000 m^2
3 156 m^2
4 2.17 mm^2
5 5.21 m/s, 54.5 m
6 1330 mm^2, 3.3 km^2
7 4.51 J
8 4667 m^2
9 4.3 m^2
10 142 m^3, 248 t
11 −2.94
12 2 V
13 152.8 V, 0 V
14 8.75 V

Exercise D3

2 (a) 2.5, 1.5; (b) 2, 2; (c) 3.0
    1.73; (d) 3, 1.27
4 (a) 22.50 cm^3, 37.5 cm^3;
    (b) 25.13 cm^3, 25.13 cm^3;
    (c) 26.97 cm^3; 46.77 cm^3;
    (d) 17.95 cm^3, 42.41 cm^3
5 (a) 17 cm^3, 25 cm^3;
    (b) 266 cm^3, 468 cm^3
6 5728 mm^3
7 16.07 mm
8 12.14 mm
9 22 mm
11 18.38 mm, 36.79 mm

Exercise D4

1 989.6 mm^2, 114.7 mm
2 110 m^2
3 53.93 mm, 68°
4 2.04 x 10^4 mm^2
5 2.36 x 10^5 mm^3
6 350 litres
7 $\pi a^2 b$
8 1.131 x 10^6 mm^3
9 73 g
10 3.80 x 10^4 mm^3
11 3.17 units

Exercise D5

1 (a) 0.8870, 0.4618, 1.921,
    1.127, 2.166, 0.521
    (b) 0.9842, −0.1771, −5.558,
    1.016, −5.647, −0.180
    (c) −0.5336, −0.8457, 0.6310,
    −1.874, −1.182, 1.585
    (d) −0.4131, 0.9107, −0.4536,
    −2.421, 1.098, −2.204
2 (a) 48.85°, (b) 63.45°,
    (c) 68.20°
3 4.508
4 0.605, 1.256
5 1.500
6 (a) 0.990, (b) 0.980, (c) 1.010
7 (a) −0.6536, −1.321; (b) 0.8090,
    −1.701; (c) 0.9888, 6.692
8 13.2 m
9 1.25 rad or 71°37′
10 22.5 m^2, 1.8 rad or 103°8′
11 3.84 m^2
12 190 mm^2
13 3.02 m^2

Exercise D6

1 1272 mm
2 17 m^2
3 (79, 62)
4 24 350 mm^2
5 14 110 m^2
6 53.03°
7 8.3 m, 9.3 m
8 19.71°
9 96.67°
10 79.6 km
11 283 mm^2, 22.8 mm, 25.5 mm
12 73.40°, 58.41°, 48.19°, 2.4 m^2
13 0.74 m
15 160 km^2
16 6.45 m^2
17 63 m^2, 91 m^2
18 6.41 m^2

Exercise D7

4 (a) $100°$ or 1.75 rad and 1,
   (b) $57.3°$ or 1 rad and $\sqrt{3}$
   (c) $60°$ or $\pi/3$ rad and $\frac{1}{2}$
5 Twice

7 2.15 and $112°$
8 3.61 and $326.3°$ or $-33.7°$
9 2.69, $68.2°$

Exercise D8

2 $\tan x = 8.415$
   $\operatorname{cosec} x = 1.007$,
   $\sec^2 x = 71.8$
4 (a) $\tan^2 \theta$, (b) $\pm\cot x$
6 (a) $-336.42°, -203.58°,$
      $23.58°, 156.42°$
   (b) $-210.68°, -149.32°,$
      $149.32°, 210.68°$
   (c) $-291.96°, -111.96°,$
      $68.04°, 248.04°$
7 $48.13°, 131.87°$

8 $-165.75°, 14.25°$
9 3.56, 5.87
10 $-139.31°, -40.69°, 40.69°,$
   $139.31°$
11 $51.67°, 128.33°, 231.67°,$
   $308.33°$
12 8
13 $60°, 70.53°, 289.47°, 300°$
14 $85.24°, 116.57°$
15 $-170.00°, -10.00°, 39.32°,$
   $140.68°$

Exercise D9

4 1.94 N s, 1.88 N s.
5 10.71 m/s^2, -3.27 m/s^2
6 $10.63, 35.04°$
7 $1.01x$ kN, $80.54°$
8 3 and -2

9 67.69 N, 396.3 N
10 $15.5, 21.2°$
11 $2693\,\Omega, 21.80°$
12 13.16 N, $74.29°$

Exercise E1

1 (c), (d), and (e) are continuous;
   (a), (b), and (f) are discrete.
7 $200.45-201.45$ mm and
   $201.45-202.45$ mm
8 $0.435-0.495$ A
9 32.117 mm
12 15
13 41, 60

14 48.8 marks
15 10.5%, 1.56%
16 44.6 h, 1.9 h
17 £60.80, £7.52
18 12, 2; 13, 2
19 24 MPa, 25.3 MPa, 2.9 MPa

Exercise E2

1 83.1%
5 £60.61
6 62
7 7.25 kN

8 54.5 cm, 5.1 cm
9 Negative, smaller
10 (a) c, (b) c
11 81.85%, 83.99%, 15.73%

Exercise E3

2 (a)
3 24
4 0.43
5 (a) and (b)
6 0.667
7 0.389

8 0.85
9 0.028
10 0.799
11 0.4
12 0.09
13 0.067
14 0.275

Answers to revision exercises

Revision exercise A1

1	11.138	9	43.66°
2	−0.81	10	2.533
3	18.99	11	2.849
4	0.116	12	13.64
5	0.794	13	0.3916
6	0.958	14	66.57
7	0.5318	15	19.84
8	2.322		

Revision exercise A2

1	(a) 5.72	2	4.7
	(b) 31.96	3	0.135

Revision exercise B1

1	(a) 0.0143	3	(a) 1032
	(b) $L = bdt^2 Y - \frac{3}{8}Mg$		(b) $V = \sqrt{(DFg/2fL)}$
2	(a) 0.30	4	39.0
	(b) $m = (A^2Pt - Abv)/v^2$		

Revision exercise B2

1	4 volts	3	6.25 litres
2	1280 kg		

Revision exercise B3

1	11.73, 0.27	3	935 mm
2	50 or −12.4	4	0.10

Revision exercise B4

1	(a) $8x^6$	2	(a) 1, (b) 3, (c) −0.143
	(b) $4a$	3	(a) ±2.24, (b) 2 or −0.5,
	(c) $4b^2$		(c) 0.618 or −1.62
	(d) $1.5c^2$	4	$x = 4, y = 2$
	(e) $4d^2$		

Revision exercise B5

1	1.2 m/s, 2.3 m/s^2	3	−1.97, 0.11, 44.88
2	1.7×10^{-5}/°C	4	0.85Ω

Revision exercise B6
1   16.1, 4.3
2   2.82

3   5.5, 1.95

Revision exercise B7
1   $x = 1.04, y = 7.57$
2   $-1, 8, 1.5, 5$
3   $x = -6.72, y = -13.28$
4   $x = 1.27, y = 3.15$

5   (a) $-2.55, 0.81$
    (b) $0.35, -2.09$
6   $-1.70, 1.19, 3.11$

Revision exercise B8
1   2.20, 0.15

2   0.0035, 1.80

Revision exercise B9
1   (a) $(7.12, 24.04°)$,
    (b) $(11.63, 101.41°)$,
    (c) $(4.06, -24.16°)$,
    (d) $(34.65, -145.35°)$

2   (a) $(10.00, 7.51)$,
    (b) $(3.42, 6.56)$,
    (c) $(9.40, 16.28)$,
    (d) $(21.34, -24.12)$
5   69.5 km

Revision exercise B10
1   (a) ON
    (b) OFF
    (c) OFF

2   (a) 1
    (b) 1
    (c) 0

3

A	B	$A + B$	$\bar{A}.\bar{B}$	$A + \bar{B}$
0	0	0	1	1
0	1	1	0	0
1	0	1	0	1
1	1	1	0	1

6

A	B	$\bar{A}$	$\bar{A} + B$	$A.(\bar{A} + B)$
0	0	1	1	0
0	1	1	1	0
1	0	0	0	0
1	1	0	1	1

Revision exercise C1
1   0.3

2   $(1, 9)$ and $(3, 5)$

Revision exercise C2
1   (a) $8x^{3/2}/3 + C$
    (b) $v^{2.4}/2.4 + C$
    (c) $-z^{-1.3}/1.3 + C$
2   (a) 84.4, (b) 1.91

3   74.7
4   4
5   $\pi/6$

Revision exercises D1
1   14 mm
2   16%
3   3.432 m^3
4   5.47 m^3
5   0.0586 m^3

6   $18\,850 \text{ mm}^3$
7   848.5 mm^2
8   $6.515 \times 10^4 \text{ mm}^3$
9   $6.379 \times 10^6 \text{ mm}^3$

Revision exercise D2
2   155 sq. units
3   25.6 mm

4   7.70×10^4 mm^2

Revision exercise D3
1   (a)   944 mm^3, 1496 mm^3
     (b)   1106.6 mm^3, 1936.6 mm^3

2   1.41 cm
3   4.46 mm down

Revision exercise D4
1   6.75×10^7 mm^3 or 0.0675 m^3
2   1194 m, 1.83×10^6 mm^3

3   70.5 mm^3
4   451 m^3

Revision exercise D5
4   15.12
5   −7.40

6   1.42

Revision exercise D6
1   171 mm
2   79.21 mm^2
3   13.16°

4   272.9 mm^2
5   2.12 km
6   3.69 m^2

Revision exercise D7
3   5, 126.87° or 2.2143 rad

Revision exercise D8
3   13.30°, 166.70°
4   −5.125, −1.158, 1.158, 5.125 rad

5   29.31°, 150.69°
6   −115.90°, −64.10°, 33.77°, 146.23°

Revision exercise D9
1   488 km/h at 010.4°
2   9.10 N at 44.9° to AB

3   9 knots

Revision exercise E1
1   9.85 kg, 1.17 kg
2   252 s, 19.3 s

3   15.25, 1.686

Revision exercise E2
1   57.9 g
2   12.7 mm
3   9

4   Higher
5   0.67σ

Revision exercise E3
1   (a)
2   14
3   0.32

4   (a) 0.91, (b) 0.21
5   0.25
6   0.23

Index